D0065106

Hubbert's Peak

Hubbert's Peak

THE IMPENDING WORLD OIL SHORTAGE

KENNETH S. DEFFEYES

Princeton University Press
Princeton and Oxford

Library of Congress Cataloging-in-Publication Data

Deffeyes, Kenneth S.
Hubbert's peak : the impending world oil shortage / Kenneth S. Deffeyes.
p.   cm.
Includes bibliographical references and index.
ISBN 0-691-09086-6 (acid-free paper)
1. Petroleum reserves.   2. Petroleum industry and trade.   3. Petroleum reserves—Forecasting.   4. Petroleum industry and trade—Forecasting.   I. Title.
TN870 .D37   2001
333.8′23211—dc21

2001032103

British Library Cataloging-in-Publication Data is available

This book has been composed in Stone Serif and Stone Sans by
Princeton Editorial Associates, Inc., Scottsdale, Arizona

Printed on acid-free paper. ♾

www.pup.princeton.edu

Printed in the United States of America

5 7 9 10 8 6 4

For Emma

It is difficult for a single author to discuss an entire industry as complex as the international petroleum industry. Although the author has made a good faith effort to evaluate the likely future course of the industry, circumstances may change and the author may have overlooked something that will turn out to be important later. Anyone considering an economic strategy should consult with a professional investment adviser. Your professional adviser—you know, the guy who assured you that dot.coms did not have to show a profit.

CONTENTS

ACKNOWLEDGMENTS

My thanks go to those who read various early versions of the manuscript. Morris Adelman, Robert Deffeyes, Robert Kaufmann, Craig Van Kirk, Jason Phipps Morgan, and Robert Solow supplied written comments, most of them quite detailed. Readers who returned marked-up manuscripts are particularly appreciated: Stephen Deffeyes, Sarah Domingo, Immanuel Lichtenstein, and W. Jason Morgan. One of Morgan's suggestions completely transformed chapter 8.

The geology librarians at Princeton University have been unfailingly helpful, even while their library is progressively being disassembled.

The editorial guidance of Joe Wisnovsky has been invaluable. A wise and experienced editor is an enormous help. On everything from the purpose of the book all the way down to the semicolons, Joe invariably could supply an immediate opinion. Sins of omission and commission remain the author's sole responsibility.

Financial support for this work came from TIAA/CREF.

Hubbert's Peak

CHAPTER 1

Overview

Global oil production will probably reach a peak sometime during this decade. After the peak, the world's production of crude oil will fall, never to rise again. The world will not run out of energy, but developing alternative energy sources on a large scale will take at least 10 years. The slowdown in oil production may already be beginning; the current price fluctuations for crude oil and natural gas may be the preamble to a major crisis.

In 1956, the geologist M. King Hubbert predicted that U.S. oil production would peak in the early 1970s.[1] Almost everyone, inside and outside the oil industry, rejected Hubbert's analysis. The controversy raged until 1970, when the U.S. production of crude oil started to fall. Hubbert was right.

Around 1995, several analysts began applying Hubbert's method to world oil production, and most of them estimate that the peak year for world oil will be between 2004 and 2008. These analyses were reported in some of the most widely circulated sources: *Nature, Science,* and *Scientific American.*[2] None of our political leaders seem to be paying attention. If the predictions are correct, there will be enormous effects on the world economy. Even the poorest nations need fuel to run irrigation pumps. The industrialized nations will be bidding against one another for the dwindling oil supply. The good news is that we

M. King Hubbert (1903–89) was an American geophysicist who made important contributions to understanding fluid flow and the strength and behavior of rock bodies. Hubbert was at the Shell research lab in Houston when he made his original estimates of future oil production; he continued the work at the U.S. Geological Survey.

will put less carbon dioxide into the atmosphere. The bad news is that my pickup truck has a 25-gallon tank.

The experts are making their 2004–8 predictions by building on Hubbert's pioneering work. Hubbert made his 1956 prediction at a meeting of the American Petroleum Institute in San Antonio, where he predicted that U.S. oil production would peak in the early 1970s. He said later that the Shell Oil head office was on the phone right down to the last five minutes before the talk, asking Hubbert to withdraw his prediction. Hubbert had an exceedingly combative personality, and he went through with his announcement.

I went to work in 1958 at the Shell research lab in Houston, where Hubbert was the star of the show. He had extensive scientific accomplishments in addition to his oil prediction. His belligerence during technical arguments gave rise to a saying around the lab, "That

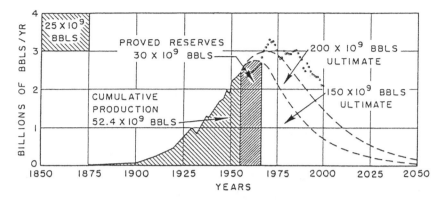

On Hubbert's original 1956 graph, the lower dashed curve on the right gives Hubbert's estimate of U.S. oil production rates if the ultimate discoverable oil beneath the curve is 150 billion barrels. The upper dashed line, for 200 billion barrels, was his famous prediction that U.S. oil production would peak in the early 1970s. The actual U.S. oil production for 1956 through 2000 is superimposed as small circles. Since 1985, the United States has produced slightly more oil than Hubbert's prediction, largely because of successes in Alaska and in the far offshore Gulf Coast.

Hubbert is a bastard, but at least he's *our* bastard." Luckily, I got off to a good start with Hubbert; he remained a good friend for the rest of his life.

Critics had many different reasons for rejecting Hubbert's oil prediction. Some were simply emotional; the oil business was highly profitable, and many people did not want to hear that the party would soon be over. A deeper reason was that many false prophets had appeared before. From 1900 onward, several of these people had divided the then known U.S. oil reserves by the annual rate of production. (Barrels of reserves divided by barrels per year gives an answer in years.) The typical answer was 10 years. Each of these forecasters started screaming that the U.S. petroleum industry would die in 10 years. They cried "wolf." During each ensuing 10 years, more oil reserves were added, and the industry actually grew instead of drying up. In 1956, many critics thought that Hubbert was yet another false prophet. Up through 1970, those who were following the story divided into pro-Hubbert and anti-Hubbert factions. One pro-Hubbert publication had the wonderful title "This Time the Wolf Really *Is* at the Door."[3]

Hubbert's 1956 analysis tried out two different educated guesses for the amount of U.S. oil that would eventually be discovered and produced by conventional means: 150 billion and 200 billion barrels. He then made plausible estimates of future oil production rates for each of the two guesses. Even the more optimistic estimate, 200 billion barrels, led to a predicted peak of U.S. oil production in the early 1970s. The actual peak year turned out to be 1970.

Today, we can do something similar for world oil production. One educated guess of ultimate world recovery, 1.8 trillion barrels, comes from a 1997 country-by-country evaluation by Colin J. Campbell, an independent oil-industry consultant.[4] In 1982, Hubbert's last published paper contained a world estimate of 2.1 trillion barrels.[5] Hubbert's 1956 method leads to a peak year of 2001 for the 1.8-trillion-barrel estimate and a peak year of 2003 or 2004 for 2.1 trillion barrels. The prediction based on 1.8 trillion barrels makes a better match to the most recent 10 years of world production.

In 1962, I became concerned that the U.S. oil business might not be healthy by the time I was scheduled to retire. I was in no mood to move to Libya. My reaction was to get a photocopy of Hubbert's raw numbers; I made my own analysis using different mathematics. In my analysis, and in Hubbert's, the domestic oil industry would be down to half its peak size by 1998. Fortunately, universities were expanding rapidly in the post-Sputnik era, and I had no trouble moving into academe.

Hubbert's prediction was fully confirmed in the spring of 1971. The announcement was made publicly, but it was almost an encoded message. The *San Francisco Chronicle* contained this one-sentence item: "The Texas Railroad Commission announced a 100 percent allowable for next month." I went home and said, "Old Hubbert was right." It still strikes me as odd that understanding the newspaper item required knowing that the Texas Railroad Commission, many years earlier, had been assigned the task of matching oil production to demand. In essence, it was a government-sanctioned cartel. Texas oil production

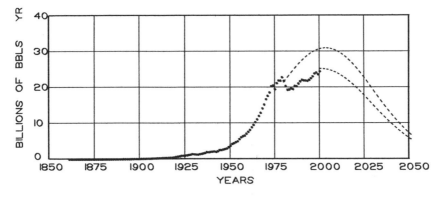

World oil production through the year 2000 is shown as heavy dots. Chapter 7 explains how Hubbert's methods were used to estimate the most likely future production. The dashed lines on the right show the probable production rates if the ultimate discoverable oil is 1.8 trillion barrels (the area under the lower curve) or 2.1 trillion barrels (upper curve).

so dominated the industry that regulating each Texas oil well to a percentage of its capacity was enough to maintain oil prices. The Organization of Petroleum Exporting Countries (OPEC) was modeled after the Texas Railroad Commission.[6] Just substitute Saudi Arabia for Texas.

With Texas, and every other state, producing at full capacity from 1971 onward, the United States had no way to increase production in an emergency. During the first Middle East oil crisis in 1967, it was possible to open up the valves in Ward and Winkler Counties in west Texas and partially make up for lost imports. Since 1971, we have been dependent on OPEC.

After his prediction was confirmed, Hubbert became something of a folk hero for conservationists. In contrast to the hundreds of millions of years it took for the world's oil endowment to accumulate, most of the oil is being produced in 100 years. The short bump of oil exploitation on the geologic time line became known as "Hubbert's peak."

In chapter 7, I explain how Hubbert used oil production and oil reserves to predict the future. We scientists don't like to admit it, but we often guess at the answer and then gather up some numbers to sup-

The 100-year period when most of the world's oil will be produced is known as "Hubbert's peak." On this scale, the geologic time needed to form the oil resources can be visualized by extending the line five miles to the left.

port the guess. A certain level of honesty is required; if the numbers do not justify my guess, I don't fake the numbers. I generate another guess. Hubbert's oil prediction was just barely within the envelope of acceptable scientific methods. It was as much an inspired guess as it was hard-core science.

This cautionary note is needed here: in the late 1980s there were huge and abrupt increases in the announced oil reserves for several OPEC nations.[7] Oil reserves are a vital ingredient in Hubbert's analysis. Earlier, each OPEC nation was assigned a share of the oil market based on the country's annual production capacity. OPEC changed the rule in the 1980s to consider also the oil reserves of each country. Most OPEC countries promptly increased their reserve estimates. These increases are not necessarily wrong; they are not necessarily fraudulent. "Reserves" exist in the eye of the beholder.

Oil reserves are defined as future production, using existing technology, from wells that have already been drilled (not to be confused with the U.S. "strategic petroleum reserve," which is a storage facility for oil that has already been produced). Typically, young petroleum engineers unconsciously tend to underestimate reserves. It's a lot more fun to go into the boss's office next year and announce that there is actually a little *more* oil than last year's estimate. Engineers who have to downsize their previous reserve estimates are the first to leave in the next corporate downsizing.

The abrupt increase in announced OPEC reserves in the late 1980s was probably a mixture of updating old underestimates and some wishful thinking. A Hubbert prediction requires inserting some hard, cold reserve numbers into the calculation. The warm fuzzy numbers from OPEC probably give an overly optimistic view of future oil production. So who is supposed to know?

A firm in Geneva, Switzerland, called Petroconsultants, maintained a huge private database. One long-standing rumor said that the U.S. Central Intelligence Agency was Petroconsultants' largest client. I would hope that between them, the CIA and Petroconsultants had inside information on the real OPEC reserves. This much is known: the loudest warnings about the predicted peak of world oil production came from Petroconsultants.[8] My guess is that they were using data not available to the rest of us.

A permanent and irreversible decline in world oil production would have both economic and psychological effects. So who is paying attention? The news media tell us that the recent increases in energy prices are caused by an assortment of regulations, taxes, and distribution problems. During the election campaign of 2000, none of the presidential candidates told us that the sky was about to fall. The public attention to the predicted oil shortfall is essentially zero.

In private, the OPEC oil ministers probably know about the articles in *Science, Nature,* and *Scientific American.* Detailed articles, with contrasting opinions, have been published frequently in the *Oil and Gas Journal.*[9] Crude oil prices have doubled in the past year. I suspect that OPEC knows that a global oil shortage may be only a few years away. The OPEC countries can trickle out just enough oil to keep the world economies functioning until that glorious day when they can market their remaining oil at mind-boggling prices.

It is not clear whether the major oil companies are facing up to the problem. Most of them display a business-as-usual facade. My limited attempts at spying turned up nothing useful. A company taking the 2004–8 hypothesis seriously would be willing to pay top dollar for existing oil fields. There does not seem to be an orgy of reserve acquisitions in progress.

Internally, the oil industry has an unusual psychology. Exploring for oil is an inherently discouraging activity. Nine out of 10 exploration wells are dry holes. Only one in a hundred exploration wells discovers an important oil field. Darwinian selection is involved: only the incurable optimists stay. They tell each other stories about a Texas county that started with 30 dry holes yet the next well was a major discovery. "Never is heard a discouraging word." A permanent drop in world oil production beginning in this decade is definitely a discouraging word.

Is there any way out? Is there some way the crisis could be averted?

New Technology. One of the responses in the 1980s was to ask for a double helping of new technology. Here is the problem: before 1995 (when the dot.com era began), the oil industry earned a higher rate of return on invested capital than any other industry. When oil companies tried to use some of their earnings to diversify, they discovered that everything else was less profitable than oil. Their only investment option was doing research to make their own exploration and production operations even more profitable. Billions of dollars went into petroleum technology development, and much of the work was successful. That makes it difficult to ask today for new technology. Most of those wheels have already been invented.

Drill Deeper. The next chapter of this book explains that there is an "oil window" that depends on subsurface temperatures. The rule of thumb says that temperatures 7,500 feet down are hot enough to "crack" organic-rich sediments into oil molecules. However, beyond 15,000 feet the rocks are so hot that the oil molecules are further cracked into natural gas. The range from 7,000 to 15,000 feet is called the "oil window." If you drill deeper than 15,000 feet, you can find natural gas but little oil. Drilling rigs capable of penetrating to 15,000 feet became available in 1938.

Drill Someplace New. Geologists have gone to the ends of the Earth in their search for oil. The only rock outcrops in the jungle are in the banks of rivers and streams; geologists waded up the streams picking leeches off their legs. A typical field geologist's comment about jungle, desert, or tundra was: "She's medium-tough country." As an example,

This 1940s rig could drill through to the bottom of the oil window. Derricks like this, although rarely used after 1950, are still a visual metaphor for the oil industry. © Bettmann/CORBIS.

at the very northernmost tip of Alaska, at Point Barrow, the United States set up Naval Petroleum Reserve #4 in 1923.[10] As early as 1923, somebody knew that the Arctic Slope of Alaska would be a major oil producer.

Today, about the only promising petroleum province that remains unexplored is part of the South China Sea, where exploration has been delayed by a political problem. International law divides oil ownership at sea along lines halfway between the adjacent coastlines. A valid claim to an island in the ocean pushes the boundary out to halfway between the island and the farther coast. It apparently does not matter whether the island is just a protruding rock with every third wave washing over the rock. Ownership of that rock can confer title to billions of barrels of oil. You guessed it: several islands stick up in the middle of the South China Sea, and the drilling rights are claimed by six different countries. Although the South China Sea is an attractive prospect, there is little likelihood that it is another Middle East.

Speed Up Exploration. It takes a minimum of 10 years to go from a cold start on a new province to delivery of the first oil. One of the legendary oil finders, Hollis Hedberg, explained it in terms of "the story." When you start out in a new area, you want to know whether the oil is trapped in folds, in reefs, in sand lenses, or along faults. You want to know which are the good reservoir rocks and which are the good cap rocks. The answers to those questions are "the story." After you spend a few years in exploration work and drilling holes, you figure out "the story." For instance, the oil is in fossil patch reefs. Then pow, pow, pow—you bring in discovery after discovery in patch reefs. Even then, there are development wells to drill and pipelines to install. It works, but it takes 10 years. Nothing we initiate now will produce significant oil before the 2004–8 shortage begins.

To summarize: it looks as if an unprecedented crisis is just over the horizon. There will be chaos in the oil industry, in governments, and in national economies. Even if governments and industries were to recognize the problems, it is too late to reverse the trend. Oil production is going to shrink. In an earlier, politically incorrect era the scene would be described as a "Chinese fire drill."

What will happen to the rest of us? In a sense, the oil crises of the 1970s and 1980s were a laboratory test. We were the lab rats in that experiment. Gasoline was rationed both by price and by the inconvenience of long lines at the gas stations. The increased price of gasoline and diesel fuel raised the cost of transporting food to the grocery store. We were told that 90 percent of an Iowa corn farmer's costs were, directly and indirectly, fossil fuel costs. As price rises rippled through the economy, there were many unpleasant disruptions.

Everyone was affected. One might guess that professors at Ivy League universities would be highly insulated from the rough-and-tumble world. I taught at Princeton from 1967 to 1997; faculty morale was at its lowest in the years around 1980. Inflation was raising the cost of living far faster than salaries increased. Many of us lived in university-owned apartments, and the university was raising our apartment rents in step with an imaginary outside "market" price. Our real standard of living went progressively lower for several years in a row. That was life (with tenure) inside the sheltered ivory tower; outside it was much tougher.

What should we do? Doing nothing is essentially betting against Hubbert. Ignoring the problem is equivalent to wagering that world oil production will continue to increase forever. My recommendation is for us to bet that the prediction is roughly correct. Planning for increased energy conservation and designing alternative energy sources should begin now to make good use of the few years before the crisis actually happens.

One possible stance, which I am not taking, says that we are despoiling the Earth, raping the resources, fouling the air, and that we should eat only organic food and ride bicycles. Guilt feelings will not prevent the chaos that threatens us. I ride a bicycle and walk a lot, but I confess that part of my motivation is the miserable parking situation in Princeton. Organic farming can feed only a small part of the world population; the global supply of cow dung is limited. A better civilization is not likely to arise spontaneously out of a pile of guilty consciences. We need to face the problem cheerfully and try to cope with it in a way that minimizes problems in the future.

The substance of this book is an explanation of the origin, exploration, production, and marketing of oil. This background about the industry is important because it sets geologic constraints on our future options. I describe the strengths and weaknesses of Hubbert's prediction methods and end with some suggestions about preparing for the inevitable. My intention is to give the reader some expertise in evaluating the problems. The experts' scenario for 2004–8 reads like the opening passage of a horror movie. You have to make up your own mind about whether to accept their scary account.

My own opinion is that the peak in world oil production may even occur before 2004. What happens if I am wrong? I would be delighted to be proved wrong. It would mean that we have a few additional years to reduce our consumption of crude oil. However, it would take a lot of unexpectedly good news to postpone the peak to 2010. My message would remain much the same: crude oil is much too valuable to be burned as a fuel.

Stephen Jay Gould is fond of pointing out that we all have difficulty rising above our cultural biases. ("All" in that sentence includes Gould.) It helps if I identify the roots of my biases. Here is my confession:

I was born in the middle of the Oklahoma City oil field. I grew up in the oil patch. My father, J. A. "Dee" Deffeyes, was a pioneering petroleum engineer. In those days, companies moved employees around wherever they were needed. I went to nine different grade schools getting through the first eight grades. During high school and college, each summer I had a different job in the oil industry: laboratory assistant, pipeyard worker, roustabout, seismic crew.

After I graduated from the Colorado School of Mines, I worked for the exploration department of Shell Oil. Right at the end of the Korean War, everybody my age got drafted. There weren't many of us. I was one of the few born right in the pit bottom of the Great Depression. I wanted to have my revenge on the army by using up my G.I. Bill at the most expensive school I could find. The geology department at Princeton turned out to be fabulous.

After graduate school, I was delighted to be asked to rejoin Shell at its research lab in Houston. Scientific progress happened very rapidly at the Shell lab. Jerry Wasserburg of Cal Tech, not known for passing out compliments freely, said that the Shell research lab in that era was the best Earth science research organization in the world. As I mentioned, it was Hubbert's prediction that caused me to get out of the oil business.

I taught briefly at Minnesota and Oregon State, and in 1967 I joined the Princeton faculty. In addition to teaching, I had the pleasure of cooperating with John McPhee as he wrote his books on geology.[11] The "oil boom" of the 1970s and early 1980s gave me a chance to participate in the industry again. As a consultant, I advised programs that drilled for natural gas across western New York and northern Pennsylvania. My programs drilled 100 successful gas wells without a dry hole; one of them was the largest gas well in the history of New York State.[12] I also served as an expert witness in oil litigation.

You don't outgrow your roots. As I drive by those smelly refineries on the New Jersey Turnpike, I want to roll the windows down and inhale deeply. But in all the years that I worked in the petroleum industry, I never came to identify with the management. I'm a worker bee, not a drone.

A couple of years ago, I was testing a new treatment on an oil well in northern Pennsylvania. I picked up a pipe wrench with a 36-inch handle and helped revise the plumbing around the wellhead. I was home again; I loved it.

CHAPTER 2

The Origin of Oil

In 1970, several major U.S. oil companies paid the government millions of dollars for oil-drilling rights off the coasts of Oregon and Washington.[1] They drilled three holes, then rode off into the sunset and never returned. They kissed millions of dollars goodbye. What happened?

1 They forgot the story about the Texas county that produced oil only after 30 initial dry holes were drilled.
2 They did not listen to the economists telling them that the amount of oil discovered depends on the number of dollars spent in the search.
3 Environmentalists were better organized in Oregon and Washington than anywhere else.
4 There was really bad news in those three holes.
5 None of the above.

You probably guessed that I would choose answer 4, although you could argue for number 5. There are geologic reasons that almost guarantee further exploration will be a waste of money and effort. What news could be so bad that major companies would write off their investment and sail away forever? It has to do with the source of oil: the rocks and their temperature history.

Early oil geologists knew that there was a little bit of organic matter in most rocks and that some of that organic material resembled oil. They speculated that some process might sweep up the organic material from a large block of sediment and move it into a concentrated spot to form an underground oil reservoir.

That gather-it-up idea started to change after George Philippi, of the Shell research lab, published a short announcement in 1957 and a longer paper in 1965.[2] Philippi reached his conclusion using instruments that seem primitive today. Essentially he smashed oil molecules into pieces, analyzed the fragments, and reconstructed molecules that would yield those pieces. It was comparable to figuring out how a computer works by breaking up the computer with a sledgehammer and examining the fragments. Not long after Philippi's announcement, an improved instrument became available called a gas-liquid chromatograph, or GLC for short. The GLC analyzed whole molecules, quickly and easily. A modern-day GLC sits on a tabletop and costs about $7,000.

The GLC outputs are graphs on a piece of paper. The graphs can be matched up like fingerprints. No heavy lifting, no heavy science, just glom the pictures. Similar GLC graphs mean similar chemistry. You can travel around Wyoming, collecting a teaspoon of crude oil from every oil well. A teaspoon is a *huge* sample for a GLC machine. Run each sample through the GLC on your tabletop, and you find out that the samples fall into two groups. There are only two fingerprints at these crime scenes. Already, you know that there are two distinctive types of oil in Wyoming.

The next question is: Whose fingerprints are they? Now you have to go to the rocks. Rock chips brought up from drilling wells are often saved. Any rock that has much organic matter in it will be chocolate brown or black. You grind up a chip with a hand mortar and pestle, along with a few drops of a solvent. Obviously, you use an organic solvent that does not occur in oil. Acetone will do. You let the solid part of the rock settle out, and then you inject the solvent, and anything that dissolved from the rock chip, into your trusty GLC machine.

After running enough chips, you have your two suspects: two rocks whose GLC fingerprints match the two oils.

The surprise is that the two rocks are rather thin layers. In Wyoming, the total stack of sedimentary rocks is about 20,000 feet thick. Each of the two layers that match the oils is about 30 feet thick. So who are these suspects?

One is a sedimentary rock about 280 million years old. It is a thin layer, a part of the thicker Phosphoria Formation.[3] (The name comes from enormous reserves of phosphate rock, mined for fertilizer, in the same formation. In the United States, formations are named only for geographic places. The U.S. Geological Survey [USGS] used a slightly unethical trick: it named an obscure gully Phosphoria Gulch, then named the formation after the gulch.) Oil now trapped in rocks both above and below the Phosphoria Formation bears the fingerprint of an origin within the Phosphoria.

The second layer is comparatively young, only 90 million years old, and consists of sedimentary beds within the Mowry Formation.[4] (No cheating on names this time; Mowry Creek is in north-central Wyoming.) The zone carrying the guilty fingerprint is no more than 50 feet thick; it is within a stack of black fine-grained rocks thousands of feet thick. Just being black or bittersweet chocolate in color is not enough. The oil is coming from a very specific zone. Usually the oil-source zones are not just a single, massive layer. Typically, there are a dozen or more thin horizons containing a lot of organic matter interspersed with less organic-rich sedimentary rocks.

The exercise with GLC patterns as fingerprints identifies the layers within the Phosphoria and Mowry as "source rocks," each matched to a specific type of oil. Go to west Texas, repeat the procedure, and you will find three source rocks. The giant oil fields of the Middle East have only two source rocks, the thickest of them less than 100 feet thick.[5] The idea that small amounts of oil are concentrated from large volumes of rock is now dead. Oil comes from very specific layers of source rocks. No source rocks, no oil.

So we have tracked the crude oil back to a few layers, tens of feet thick, within stacks of sedimentary rocks more than 20,000 feet thick.

This core from a well in the Paradox Basin (southeastern Utah) is a typically black source rock for oil. The Paradox Basin contains an abundance of source rocks; some informal estimates indicate that half the subsurface pore space in the basin is filled with oil.

What is special about those thin layers? Sometimes, as in Wyoming, there are thousands of feet of adjacent sediment layers that are black with organic matter, but those thick beds don't leave their fingerprints on crude oil. An important key is the *amount* of organic matter. Consider a sediment deposited containing 1 percent of cells from dead algae. If that sediment is buried beyond 7,500 feet, the dead algae will start to crack to form oil. But what do we have? A rock made of clay and water with a few scattered drops of oil. In the early days, petroleum geologists knew that this was a problem. The scattered oil drops would simply sit there. Papers were written about "mobilization" and "micellurization," but they were equivalent to whistling while walking past a graveyard at night. When a source rock containing lots of organic matter begins to crack into oil, the oil drops merge inside the rock to form continuous strings and blobs of oil. Now as the rock proceeds farther into the oil window and as molecular cracking continues, oil starts to be squeezed out into the surrounding rocks. Oil gets squeezed out both above and below. When you squeeze a kitchen sponge, water comes out both sides.

In his brief 1954 announcement, Philippi pulled a second rabbit out of his hat. The source rocks do not start out containing oil. The molecules in oil are smaller pieces cracked from larger organic molecules in the source rock. Oil refineries use the same trick.[6] They break up natural oil molecules into smaller fragments for high-grade gasoline. The tallest objects in an oil refinery are called "cat crackers," but they don't crack cats. "Cat" is short for "catalytic"; the cat cracker uses heat plus a catalyst to speed up the reaction.

In crude oil, the simplest molecules are chains of carbon atoms with hydrogen atoms strung along both sides and at the ends of the chain. Molecules made of hydrogen plus carbon are called (guess what) hydrocarbons. We Okies call the simplest chains "paraffins"; organic chemists call them "normal alkanes." They are household products. The shortest, with only one carbon (methane), is the dominant ingredient in natural gas. Chains with three or four carbons (propane and butane) are the bottle gas in the backyard grill or the camper trailer. The 8-carbon chain is octane, made famous by gasoline ratings. Lubricating oils have around 15 carbons in the chain. The 30-carbon paraffins are used to seal jars of homemade jelly and to coat waxed paper. The paraffin in the jelly jars gives its name to the whole series. Although chains thousands of carbons long do not occur in nature, petroleum refiners assemble ultralong chains to make the familiar plastic polyethylene. Those floppy plastic bags from the grocery store are polyethylene.

Although the hydrocarbons produced from wells are usually listed as "oil" and "gas," there is an in-between product that sells for high prices. Hydrocarbons with three to five carbon atoms go under a bunch of names: gas condensate, natural gas liquids, drip gas, and "white gold." Some oil field workers pour the stuff directly into the gasoline tanks of their pickup trucks, a practice that is both dangerous and illegal. (In 1931, my father drove the round trip from Oklahoma City to El Paso for his annual duty as a U.S. Army Reserve officer using

The first five members of the paraffin family contain from one to five carbon atoms. The molecule at the top, with one carbon (black) and four hydrogen atoms (white), is known as "methane" and is the dominant component in natural gas. The molecules with three and four carbon atoms are named "propane" and "butane," respectively, and they are the bottle gas in backyard grills and on camper trailers.

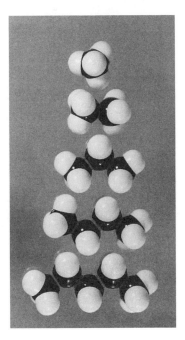

a 55-gallon drum of gas condensate. He used his mileage reimbursement to pay the doctor bill when I was born.)

Hydrocarbon chains are produced naturally by a wide assortment of plants and animals. Our ear wax is a mundane example. Hydrocarbons hate water; they do not mix with water; hydrocarbons repel water. Among chemists, "hydrophobic" is not a synonym for rabies. Organisms, including people, are fundamentally bags of water. The best way of structuring a water-rich system is to use the contrast between water-hating (hydrophobic) components and water-loving (hydrophilic) entities. At one end of the scale, simple single-celled marine algae use hydrocarbon molecules to create a stable cell wall. At the other end, the myelin coatings that serve as electrical insulation around nerve cells in the human brain are also hydrocarbon chains.[7]

The hydrocarbons made by organisms, from algae to humans, have a common oddity—literally an oddity: most of them contain an odd number of carbon atoms. (Don't ask me why; check with your neighborhood biochemist.) A pattern from a GLC machine spreads

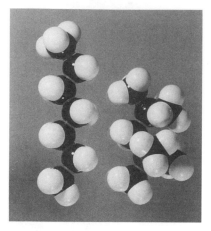

In addition to simple straight chains, the paraffin family also contains chains with side branches. The simple unbranched chains are more susceptible to pinging, an uncontrolled detonation in automobile engines. These two molecules are the basis for the octane rating posted on the gasoline pump. The seven-carbon straight molecule on the left is given an octane rating of 0; the branched paraffin on the right is the standard for 100 octane. Real gasolines are rated by comparison to mixtures of these two molecules. Straight chains are easily digested by bacteria. Biodegradable detergents are manufactured from unbranched paraffins.

out paraffin molecules, with the shortest chains on one side of the paper and the longest chains on the other. Put in almost any organic material and you get an alternating pattern: large amounts of the odd-numbered hydrocarbons and much smaller amounts with even numbers. In the oil business, it is called the "odd-even" predominance.

Philippi's second observation was that the odd-even predominance is completely missing in crude oils. Even numbers of carbons are just as abundant as odd ones. All potential sources from organisms contain mostly odd-numbered chains. However, if those original chains were broken apart at random places in between the original carbons, the original odd-even predominance would be eliminated.

Even after Philippi's observations were published in a longer paper, not everyone paid attention. When I was a new faculty member at Princeton, I attended a big public lecture entitled "Organic Matter through the Ages." The speaker was a Berkeley professor who won the Nobel Prize for discovering how photosynthesis worked. He showed slides, most of them GLC output, to illustrate natural organic materials of different ages. His very last slide was a GLC result from a meteorite that had recently landed at Pueblo de Allende, in Mexico. His

GLC pattern showed a strong predominance of odd over even carbon chains! As the lecture ended, my pulse rate must have been up around 150. I was going to ask the first question: "Could we please have that last slide back? Did you notice, sir, that your meteorite sample was contaminated by burro droppings?" Fortunately, the person who had introduced the speaker stood up and said, "Thank you for a wonderful lecture. Good night, everyone." Some months after that lecture, I ran across a member of the speaker's laboratory staff at a meeting. I asked about the odd-even pattern from the Allende meteorite. He said that they did realize it was contamination, but The Great Man had come barreling through his labs saying, "Gimme the latest stuff. I gotta go give a lecture at Princeton."

What does it take to break up the original hydrocarbons? Time and temperature. Chemical reactions run faster when it is hot. The required time and temperature can be determined either in the ground or in the lab. (We'll come back to the lab part later in this chapter.)

A sedimentary layer with plenty of organic matter will show its original predominance of odd over even numbers of carbons as long as it has never been hot enough to run the cracking process. This type of rock is referred to as an "immature" source rock. In contrast, if an organic-rich sedimentary rock has ever, in its long history, been hotter than the cracking temperature, then the rock will contain oil with equal numbers of odd and even carbon chains: a "mature" source rock.[8]

A drill hole that goes deep into the ground usually finds that the temperature increases by about 14°F per 1,000 feet (25°C per kilometer). The boundary between immature and mature source rocks is found at roughly 7,500 feet depth, where the temperature is around 180°F (82°C). John McPhee likened this to "the temperature of a cup of coffee."[9] A million years at coffeepot temperatures will break hydrocarbon chains.

A second observation from deep drill holes: deeper than 15,000 feet, you usually do not find oil. You can roast a turkey at those temperatures: 295°F (145°C). Given geologic time, the temperatures at 15,000 feet will break *all* the carbon-carbon bonds. (There is a minor exception: with really rapid burial some oil persists to 17,000 feet.)[10] What is left are molecules with only one carbon, methane, the dominant component of natural gas.

In the oil fields, as elsewhere, people love a good rule of thumb. The range from 7,500 to 15,000 feet became known as the "oil window." Source rocks had to be buried at least 7,500 feet to start generating oil. However, burial beyond 15,000 feet would crack all the oil down to natural gas. But wait a minute! The first American oil well, the Drake well of 1859, was only 75 feet deep. Two explanations: (1) after the source rock has been buried beyond 7,500 feet, surface erosion can bring the whole package back near the surface, and (2) oil, once liberated from the source rock, can migrate upward to form shallow oil fields and even appear as surface oil seeps.

So now we have an agenda: an oil province has to contain organic-rich source rocks, and the source rocks have to have been buried below 7,500 feet but no deeper than 15,000 feet. If these ingredients are not present, no amount of drilling is going to find oil. Here's an example: the sediments on the floor of the deep sea are about 3,000 feet thick. Even if organic-rich layers were present near the base of the sediments, they would not even be halfway to the top of the oil window. This paragraph wipes out 60 percent of the Earth's surface as a potential source for oil.

The old-timers looked for oil seeps. Some oil seeps were subtle; some were obvious to anyone. The Los Angeles Basin contained a number of oil seeps and tar pools in addition to the famous La Brea tar pits.[11] (I never did believe the rumor about a movie starlet named La Brea Tarpitz.) When an oil company started exploring Kuwait in 1932, the Kuwaitis suggested drilling near the asphalt pit at Burgan. "Nope," the company in effect replied, we have to do our geophysical work and decide where to drill. After drilling several dry holes, the company agreed in 1938 to drill next to the asphalt lake. Burgan became the second-largest oil field in the world, 70 billion barrels.[12] Time out for World War II; Kuwait got no significant oil revenue until 1947.

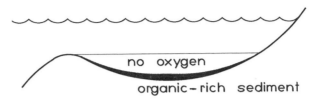

This diagram expressed a widely used hypothesis: in the absence of deep-water circulation, organic matter would be preserved because oxygen in the bottom water would be used up.

Either surface oil seeps or successful producing oil wells identified areas as "oil country." Trying to discover additional oil fields, within known oil country, is tough enough. Establishing that a new "frontier" province is oil country involves huge risks and heavy expenditures. Understanding the geologic circumstances that deposit organic-rich source rocks helps in sorting out the winners from the losers. Records in the rocks that document temperature history also help in identifying the boundaries of "oil country."

So where are organic-rich sediments deposited? The fate of most organic matter is to get oxidized after the organism dies, mostly by the action of bacteria. Robert Frost called it "the slow smokeless burning of decay."[13] The obvious place to preserve organic matter is an environment where oxygen is lacking. The standard explanation was places like the Black Sea, some of the Norwegian fjords, or the Cariaco Trough in the southern Caribbean. I published a diagram in a beginning geology textbook.[14] The idea was that deep water in a basin was isolated by a shallower spot, a "sill," named for the analogy to a doorsill. The story had a weakness. None of the present-day silled basins deposit sediments with enough organic matter to make a decent source rock. (Or at least a Shell source rock; ExxonMobil was easier to satisfy.) In one way, it was sort of expected that no source rock would be actively forming today. Source rocks make up only a thin part out of very thick piles of sediment. During most of geologic time, Wyoming (or the Middle East) was *not* depositing source rocks. It would take a lucky coincidence for one of these rare events to be going on today.

The circumstances that produced oil source rocks obviously did happen sometimes, but I could get no help from studying modern sediments. Help came as a surprise. One of my smaller tasks at Princeton was organizing Thursday lunchtime sessions for the undergraduate juniors to hear faculty talks about active research projects. It started the juniors thinking about their senior thesis projects. One Thursday, about four years ago, Professor Jorge Sarmiento was giving a talk entitled "Sapropels in the Eastern Mediterranean." I didn't actually remember what a sapropel was, but the lecture came with a free lunch. I settled down for my postlunch quiet time, if not an actual nap, and Jorge started by explaining that sapropels were highly organic-rich sediment layers. Sapropels were not forming today, but in the Eastern Mediterranean sapropels had been forming off and on in the recent geologic past. One of my eyes began opening.

He went on to explain why the organic-rich sapropels were formed intermittently. A persistent aspect in all oceans is for surface organisms to grow using nutrients dissolved in surface seawater, such as phosphate, nitrate, silica, and iron. When the organisms die, either their dead remains or the fecal pellets from their predators fall into the deep water.

Jorge pointed out two different ways that the Eastern Mediterranean Sea could exchange water with the rest of the ocean. One had surface water coming in and deeper water going out; the other was just reversed, deep water in and surface water out. Since organisms strip nutrients into the deep water, circulation of surface water in and deep water out turns the ocean locally into a nutrient desert. Snorkelers and divers love nutrient deserts; the water is clear blue, and the bottom is visible 60 feet down.

The alternative, with deep water coming in and surface water going out, traps nutrients. Some dissolved nutrients come in with the deep water, but almost no nutrients leave with the surface water. As the nutrients built up, the Eastern Mediterranean turned into a highly fertilized garden. The abundant rain of dead bodies and fecal pellets

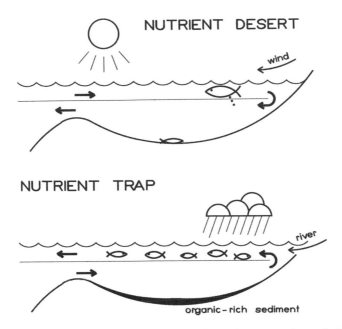

NUTRIENT DESERT

NUTRIENT TRAP

organic-rich sediment

Dead organisms and fecal pellets transfer biological nutrients from shallow water to deep water. Circulation in an arm of the sea depends on the balance between surface evaporation and freshwater input. If evaporation exceeds the rain and river input, then the nutrient-rich deep water is exported. When freshwater input is greater than evaporation, biological nutrients are retained in the basin, and high organic productivity causes organic-rich sediments to be deposited.

from the surface overwhelmed the limited oxygen content of the bottom water. Organic-rich sapropels were deposited. At this point I was wide awake. This is the source of the world's oil.

Near the end of his lecture, Sarmiento explained how the circulation changed back and forth from nutrient trap to nutrient desert. Is the surface water saltier, and therefore heavier, than the water beneath? Rain and rivers running into the Eastern Mediterranean tend to make the surface water less salty and lighter. On the other hand, evaporation off the sea surface, promoted by sun and wind, increases the surface salinity. It is the balance between these two opposing forces that determines the circulation. There is no in between; this ballgame does not end in a tie. Often, when we subtract two big numbers to get the difference between them, the answer can be either positive or negative. I was sitting bug-eyed on the edge of my chair as Sarmiento showed sediment cores from the Eastern Mediterranean with several different organic-rich layers, sapropels. The Eastern Med is a nutrient desert today, but it was a nutrient trap several times in the past million years.

Flipping back and forth from nutrient trap to nutrient desert explains why the ancient source rocks seldom consist of one massive organic-rich layer. It's interesting to think about how the balance of fresh water versus evaporation could change. Everyone knows that weather and climate are fickle; changes in rain and wind could shift the balance. There is another factor: some rivers shift around during geologic time. Look at a map of the Mediterranean. There is no high ground to prevent the Nile from following the Ismalia Canal into the Red Sea instead of the Mediterranean. If I loaned you half a dozen bulldozers over a long weekend, you could probably flip the Nile into the Red Sea. *Don't do it!* Better you should reverse the flow in the Volga-Don Canal and turn the Eastern Mediterranean back into a nutrient trap and pull carbon dioxide out of the atmosphere.

As Sarmiento ended his lecture, I went streaking to the library. The nutrient-trap, nutrient-desert hypothesis had important advantages. It explained why oil source rocks usually were not single massive layers. The diagram in my textbook would not easily explain why ancient source rocks often fluttered off and on, like the recent history of the Eastern Med. I would have to install and uninstall that sill in my textbook diagram each time the organic content changed in the source rock. My real interest in the library was finding out whether petroleum geologists used the nutrient-trap, nutrient-desert idea. It turns out that explanations for depositing oil source rocks are almost as numerous as explanations for Dow Jones stock prices. After wading through dozens of published papers, I found one that mentioned nutrient traps and nutrient deserts.[15] Rats! I was having fantasies about using Sarmiento's insight to publish a paper on the origin of oil. The basic idea had already been published, but it looks as if changing quickly from a nutrient trap to a nutrient desert deserves wider attention.

So we need a source rock with lots of organic matter. But how much is "lots"? Here the oil companies disagreed. ExxonMobil was happy to consider rocks with as little as 2 percent organic matter as potential source rocks. Shell wouldn't look at anything less than 8 percent. Rather than labeling this disagreement as a scientific failure, let's consider it a virtue of the free-enterprise system. The company with the better interpretation will find more oil. We need the oil.

It isn't enough to have organic-rich source rocks. The package of rocks has to go through an appropriate thermal history: hot enough to form oil but not hot enough to break all the oil into natural gas. The rocks had to have passed into the "oil window." In effect, we want to ask the rock: "Are you now, or have you ever been, below 7,500 feet, and have you ever been below 15,000 feet?" Only those rocks that can answer yes to the first question and no to the second are candidates for oil exploration.

Before 1970, natural gas was a by-product of modest value, like corncobs and wheat straw. During the oil crisis of the early 1980s, prices for 1,000 cubic feet of natural gas jumped from five cents to three dollars. Suddenly, all those places that were avoided because they were below the oil window became profitable. The "oil boom" of the 1980s was primarily a gas boom.

There are several different ways to get a rock to disclose its maximum temperature of burial. In 1915, long before we knew about the oil window, David White pointed out that coal beds contain a temperature record.[16] Coal beds progress from loose peat, to soft brown lignite, through a range of bituminous coal grades, to hard shiny anthracite. White showed that oil was associated with the bituminous coal grades. Lignite was not enough, and anthracite was too much.

White was not saying that oil and coal were coming from the same source. He was telling us that within a package of oil-bearing rocks, if any coal were present, it would be bituminous coal. It turned out that you didn't have to find a whole bed of coal to make the test. Even microscopic bits of buried vegetable matter change into micro-

At higher temperatures, most of the molecules in crude oil break down into more stable forms. At extreme temperatures, the stable forms are crystalline minerals composed only of carbon: graphite and diamond. Although graphite and diamond do not form in oil, the most stable molecules resemble small portions of the graphite and diamond structure. The bottom image shows six-carbon aromatic rings linking together in the same flat hexagonal array found in graphite. The upper molecule contains a core of 10 carbon atoms arranged exactly like a fragment of the crystal structure of diamond.

scopic bits of coal. So we use a microscope. Shiny reflections from tiny bits of one variety of coal, called "vitrinite," have been calibrated to measure the maximum temperature of burial.[17]

There are thermal indicators other than coal. Many sedimentary rocks preserve pollen grains; the pollen grains get toasted darker and darker with increasing temperatures. I was surprised to learn that pollen grains survive in rocks 100 million years old. I was even more surprised at the recipe for getting pollen out of the rock. You grind up the rock, go in the chemistry lab, and pour on, one at a time, every strong acid, strong base, and oxidizing agent in the lab. Most of them would burn your skin in a fraction of a second. Everything in the rock is destroyed except the pollen. My congratulations to the plants; hay fever is forever.

In his Pulitzer Prize–winning book, *Annals of the Former World,*[18] John McPhee describes how Anita Harris of the USGS discovered that the microscopic teeth of a cryptic fossil, called "conodonts," changed from light amber color to brown to black as they were exposed to

higher and higher temperatures. Light brown conodonts are the signature of the oil window.

Conodonts heated in the laboratory change from amber to black
with increasing temperature. There is an obvious problem: geologic
times are far longer than laboratory times. I came to appreciate this
because I had an undergraduate student, Vivian Rejebian, who wanted
to go on to dental school. She planned to major in geology, but we had
to find a senior thesis topic for her. Aha! Conodonts are teeth. Anita
Harris agreed to supervise Vivian on a summer research project. It was
a medium-tough project.

Fifteen years earlier, Anita had calibrated the times and temperatures that would change the colors of conodonts, but she did the
heating in the lab air. Anything, like water or carbon dioxide, given
off during the heating was free to leave. The work needed to be redone,
with the conodonts sealed along with water and heated inside high-
pressure bombs. Anything that could go wrong in that experiment did
go wrong. But with patience, lots of telephone calls, and experienced
lab people at the USGS, Vivian finished the summer with usable data.
Her senior thesis was published in the prestigious *Bulletin of the Geological Society of America*. Anita Harris and the lab scientists at the USGS
were joint authors, but the first author was Vivian Rejebian, George-
town School of Dentistry.[19]

From the microscopic bits of coal, pollen grains, conodonts, oil
fields, and deep drill holes, the temperature history of the outer 20,000
feet of the continents is reasonably well known. We began to realize
that neither the age of the oil source rock nor the age of the reservoir
rock was necessarily the age of the oil field. In the Wyoming example,
the Phosphoria source rock is about 280 million years old, and the underlying sandstone that forms the producing reservoir is about 300
million years old. However, sediment was still accumulating in the
basins between the Wyoming mountain ranges as late as 20 million
years ago.[20] The original tendency is to think of the oil in terms of the
280- and 300-million-year ages of the rocks involved, but the source
rock may have been pushed into the oil window only 20 million years
ago. I realize that it may sound silly to talk about "only" 20 million

years. Thinking that a million years is a short time is an occupational disease among geologists.

There is an academic puzzle alongside the practical problems. Oil source rocks ought to get progressively buried; they ought to get shoved right through the oil window. After a piece of continental crust is formed, or heavily modified, by mountain building, slow cooling of the crust and the underlying mantle is expected to make the rocks gradually cooler and therefore denser, and they ought to subside. For a while (50 million years is a "while") after mountain building, cooling and subsidence do take place. The continental surface usually drops just below sea level, and new sediments cover over the former mountain range. I would have expected the cooling to continue and all but the youngest parts of the continents to be covered by thick accumulations of sedimentary rocks. It does not happen. Continental surfaces tend to stick pretty close to sea level for at least a billion years.

Academic geologists call this the "freeboard" problem. On a ship, freeboard is the height of the deck above the waterline. (Not to be confused with "free room and board," a euphemism for a prison sentence.) If a continental surface had been uplifted for a long time, there would be high-temperature–high-pressure rocks exposed at the surface. If the continent sank to important depths, we would find extremely thick accumulations of sedimentary rocks. Neither happens. Most continents tend to hang around within 3,000 feet of sea level, and their long history seems to be a tale of accumulating a little sediment and eroding a little sediment. The continents maintain a small and nearly constant freeboard. The practical result for the oil industry is that sediments on top of the stable continents dance in and out of the oil window for geologically long times.

The probable explanation of a stable, but small, continental freeboard comes from the work of W. Jason Morgan, a Princeton professor. I get great amusement out of introducing him to present-day undergraduate students: "This is Professor Morgan; he discovered plate tectonics." Invariably, the students squint at Jason with looks that say, "Shouldn't you be dead by now?" Today's students learned about plate tectonics in grade school; they assume that the idea has been around since the Late Pleistocene. Actually, Morgan was 31 years old when he gave his first talk announcing what we now call plate tectonics.

For an encore, after announcing plate tectonics, Morgan started reconstructing past plate motions by following trails of volcanic rocks. In addition to volcanoes at plate boundaries, the Earth today has about 30 major volcanic sources that are not necessarily at plate boundaries: Yellowstone, Hawaii, Iceland, the Galapagos. Informally, they are called "hotspots." Morgan used age measurements of now dead hotspot volcanoes to track how plates moved in the past. He went on to point out that a hotspot does more than just stitch a narrow line of surface volcanoes across the surface of a plate. A swath of continent, or of ocean floor, more than 500 miles wide is uplifted as it passes by the hotspot. Whether the uplift is due to heating or to added material is not certain. A back-of-the-envelope calculation shows that any area of old continental surface will encounter these hotspot fringes one or several times during its history. This is Morgan's explanation for the approximately constant continental freeboard.[21] It also explains why rock layers near the continental surface get bounced in and out of the oil window like a dribbled basketball.

Professor Morgan's explanation has an amusing side effect. Petroleum geologists often refer to oil provinces as "basins": the Illinois Basin, the Michigan Basin, the Williston Basin. Morgan explains that we had been looking at the hole instead of the doughnut. In Morgan's view, the uplifts between basins—the Findlay arch, the Kankakee arch—are former hotspot tracks. Basins are simply the leftover spaces missed by the most recent hotspot trails.

Officially, oil source rocks and the oil window make a rather nice package, but there is a problem with the standard story. Several different laboratories have cracked oil source rocks, or toasted conodonts, with similar results. In the lab, you can keep the heat on for a long time (a few weeks) or a short time (a few seconds). However, the shortest geologic times are a few million years. An appropriate graph, called an "Arrhenius plot," allows us to extend lab times out to geologic times.[22] The slope of the line on the Arrhenius plot depends on the "activation energy." (Activation energies are usually reported in kilocalories/mole, abbreviated as kcal/mole.) Several labs agree: the measured activation energies for cracking source rocks to oil are close to 60 kcal/mole. However, when I try out 60 kcal/mole in the ground, I get an oil window that is only 1,500 feet thick instead of the real oil window. I can match the observed oil window in the ground if I choose 15 kcal/mole. If the lab measurement of 60 kcal/mole were correct, the oil window would be so thin we wouldn't have an oil industry.

A useful concept in understanding reaction rates is an "Arrhenius clock"—a clock that runs faster at higher temperatures. Not just somewhat faster, but a speedup that matches the activation energy of a chemical reaction. I think I first read about Arrhenius clocks in an Eastman Kodak advertisement about developing color film, but neither Kodak nor the genius who wrote Kodak ads for *Scientific American* could locate the ad. (I have wondered whether the wire going into a McDonalds french-fry vat is an Arrhenius clock. If it isn't, it should be.) The organic-rich source rock in the ground is an Arrhenius clock. The hotter the environment, the faster the source rock generates oil. I obtained 15 kcal/mole in the previous paragraph by writing a computer program for an Arrhenius clock and by trying different activation energies until the Arrhenius clock matched the oil window. At temperatures just under the boiling temperature of water, a reaction with an activation energy of 15 kcal/mole will double its speed for a 18°F (10°C) rise in temperature. If the reaction's activation energy is 60 kcal/mole, the same temperature rise will make the reaction run 10 times faster.

Activation energies describe more than just chemical reactions. I found printed on the side of a milk carton a list of the times that the milk would take to go sour at various temperatures: it made a beautiful straight line on an Arrhenius plot. We also know to heat up syrup if it comes out of the refrigerator too thick to pour. The viscosity of either water or syrup makes a straight line on an Arrhenius plot, but the slopes of the two lines are wildly different.

When petroleum refiners need to speed up a chemical reaction, they usually turn to a catalyst. A *catalyst* is defined as something that speeds up a reaction but is not itself consumed. A catalyst behaves much like a divorce lawyer. During a year, one divorce lawyer can assist a large number of couples splitting up, but at the end of the year the divorce lawyer is still there and still active. Also, a lawyer is supposed to be promoting only divorces that would happen anyway, sooner or later. Same with catalysts: a catalyst can speed up only those reactions that would go spontaneously, even if slowly, by themselves. If you want a demonstration of a catalyst, you need an ordinary sugar cube and some cigarette ashes. Try to light a sugar cube with a match. It won't light. Dust a sugar cube with cigarette ashes; hold it over a match, and it lights and burns up. Chemistry is fun.

Back to the problem in the ground: Could a natural catalyst lower the activation energy of source rock cracking? Petroleum refineries do this all the time, but their typical catalyst consists of platinum dispersed in a synthetic version of the mineral faujasite. Faujasite in nature is so rare that it exists in tiny amounts at only nine localities in the world.[23] Platinum is, to put it mildly, scarce. Petroleum refinery practice isn't much help. Further, if there is a natural catalyst, it has to be available everywhere. We don't have oil source rocks, either in the lab or in the ground, that require higher temperatures because they lack the natural catalyst.

If the presence or absence of a catalyst isn't the explanation of the low activation energy, what could it be? A possible explanation, which I find convincing, is that there are two utterly different chemical reactions: one in the lab and one in the ground. If the lab reaction has a larger activation energy than the underground reaction, the two lines describing the reactions will cross each other on an Arrhenius plot. For time spans that you have the patience to observe in the lab, the reaction with the larger activation energy goes faster. In the ground, the lower activation energy can be the faster of the two reactions.

It is quite clear that the lab reaction, and a similar reaction used in petroleum refineries, break a hydrocarbon chain into two fragments. One or both of the fragments have to be unstable; the hydrogen atom that would normally be at the end of hydrocarbon chain is missing. The mystery is the reaction in the ground. Whodunit? Like a

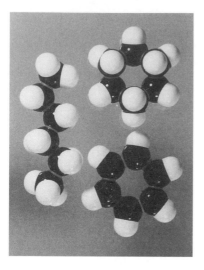

These are three fundamental types of hydrocarbon molecules in crude oil. In these models, the carbon atoms are shown as black balls, and hydrogen atoms are white. On the left is a straight chain known as a paraffin (or alkane) molecule. In the upper right is a ring called a naphthene (or cycloalkane) molecule. The ring on the lower right has a different kind of bonding, shared around a ring of carbon atoms. Most molecules with this type of bonding have distinctive odors; the family is called "aromatic." The particular molecule on the lower right, with six carbon and six hydrogen atoms, is benzene. Benzene is a major component in gasoline and is also a known cause of cancer.

good mystery writer, I have to fill you in on some clues before unveiling my candidate.

Crude oil contains, for fun and profit, lots of other molecules besides the simple hydrocarbon chains. The U.S. Bureau of Mines identified several thousand different hydrocarbons in a crude oil sample from the Brett #6 well at Ponca City, Oklahoma.[24] The oil contains branched hydrocarbons with the same chemical formula as the straight-chain hydrocarbons. There are hydrocarbon rings of two general types: naphthenic rings (cycloalkanes) and aromatic rings (benzene and its derivatives). Aromatics get their name because you can smell them. Pure paraffins and naphthenes are odorless. (The smell that warns us of leaking natural gas or bottle gas is a synthetic molecule that is added for safety.) There are three fundamental types of crude oils, depending on whether the paraffins, naphthenes, or aromatic molecules predominate. The paraffins are likely the broken fragments from larger hydrocarbons. The aromatic molecules have some natural precursors, and they are exceedingly stable molecules. Once aromatics are formed, they usually don't get modified. But what about the naphthenes?

Rings with side branches are actually more common in oil than simple rings. The smaller molecule at the top is a five-carbon ring with a "methyl" side branch composed of one carbon and three hydrogen atoms. The larger molecule below is a six-carbon ring with two methyl side branches.

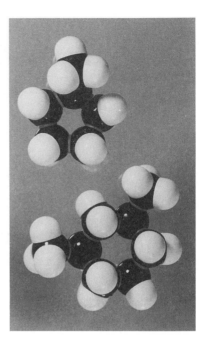

When I was in the sixth grade in Great Bend, Kansas, one of the spelling words was *naphtha*. I didn't know what it meant, and the spelling didn't seem logical; I finally memorized the spelling like a cheerleader's chant: N-A-P-H-T-H-A. (Later, the dictionary showed me why *naphtha* did not spell like a regular English word. It was a Latin word, borrowed from the Greeks, who got it from the Persians, who picked it up from the Assyrians and Babylonians.) As I pursued professional work in the Shell research lab, those ring-shaped naphthenes stuck in my mind as having no logical explanation. There were no obvious natural precursors, yet naphthenic rings made up a substantial percentage of most oils. An inspiration came from high-energy physics. In the past 10 years, there have been attempts to unify gravity with the rest of physics using "string theory." Elementary particles are explained as strings that exist in multidimensional space, but the higher dimensions are rolled up on a tiny scale where we cannot easily examine them. I didn't understand string theory, but one of their diagrams carried a message: Ohhh . . . that's where naphthenic

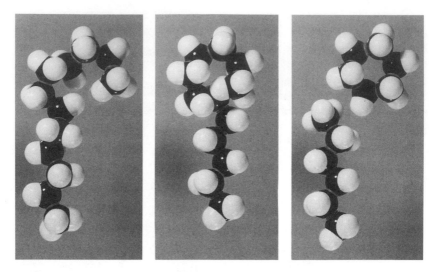

A naphthenic-ring molecule might result from the breakup of a straight chain. Although we typically draw simple chains as geometrically straight, in liquid oil the chains are flexible. The left-hand photograph shows one end of a chain bending back on itself. The center picture shows a hydrogen atom at the end of the chain approaching the bond between the sixth and seventh carbon atoms in the chain. On the right, a naphthenic six-carbon ring has separated, leaving a shorter simple chain behind.

rings came from. In a paper I published in 1982, I had modeled the breakdown of hydrocarbon chains by simply breaking the straight chain into two pieces, which is probably the reaction that happens on a laboratory time scale.[25] The trouble was that two hydrogen atoms were needed to fit onto the broken ends. There were some possible explanations, but none of the explanations fit very well with what was seen in crude oils. The model I borrowed from string theory (by enlarging the scale several billion billion times) never needed extra hydrogen atoms on the ends. Other reactions are possible. Large rings can twist into a figure eight and split into two rings. Two straight-chain hydrocarbons can cross and trade ends, which solves a second puzzle. Typical sources, such as the cell walls of marine algae, have 15 or 17 carbon atoms in their chains. Crude oils usually contain some longer chains. The crossover reaction between straight chains could make chains longer than either parent.

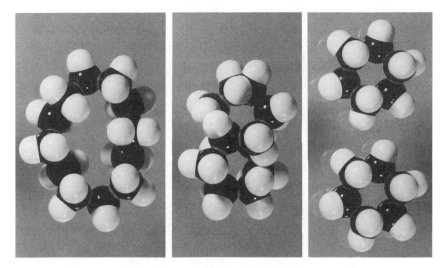

Large naphthenic rings, with more than seven carbon atoms, are rare or absent in crude oils. These models suggest how large naphthenic rings might split in two. The 12-carbon ring on the left can fold up into the figure-eight configuration shown in the center. If a pair of carbon-carbon bonds trade places, two 6-carbon rings will be formed.

I ran some computer simulations, and the computed products look rather like crude oil. Modern petroleum refineries routinely re-arrange small hydrocarbon molecules,[26] a process known as "isomerization." However, refineries are not about to waste time and energy; their rearrangement reactions utilize carefully selected catalysts.

The bottom line is that we have a puzzle: the 15 kcal/mole activation energy that is required to explain the oil window. Explaining the puzzle using the natural equivalent of a process used in oil refineries is only an interesting idea, but at least it is a start toward resolving the mystery.

So, back to the drilling off Oregon and Washington. Why would major oil companies drop several million dollars each and ride off into the sunset? I have no inside information, but is not at all difficult to guess what happened. The first two questions to be asked in a new province are (1) whether there are organic-rich source rocks and (2) whether the source rocks have been in the oil window. This

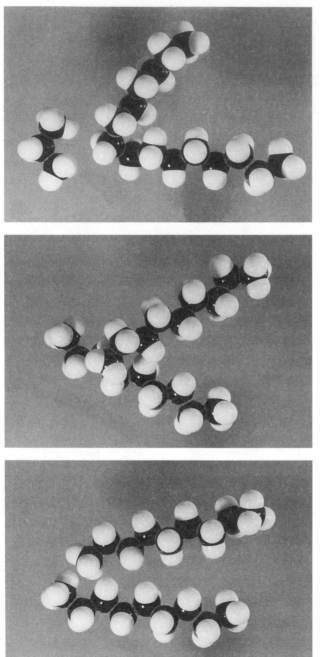

Crude oils contain some straight chains that are longer than the biologically formed chains in the source rock. On the left, two chains with 9 carbon atoms each approach each other. In the center, the end of one chain approaches the bond between the third and fourth carbon of the other chain. The right-hand image shows a short 3-carbon chain, leaving behind a 15-carbon chain.

requires a test well or two, but the test wells do not have to discover oil in order to answer the two questions. In fact, the companies often drill a hole or two in locations chosen *not* to find oil. A joint test is supposed to be called a "stratigraphic test"; around the office coffeepot it is a "strat test." Several companies share the expenses and divide up the samples from the well. Each company does its own lab work and reaches its own conclusions. If Shell walks away and ExxonMobil stays in, you know that the strat test found potential source rocks richer than 2 percent and leaner than 8 percent.

In the Oregon-Washington experience, there are two circumstances that would have caused every company to kiss the offshore area "Goodbye, goodbye forever." One would be the complete absence of source rocks in the well samples. If the Oregon-Washington shelf had always been facing out toward the open Pacific, then organic-rich source rocks might never have accumulated. The other circumstance, equally fatal, would be the presence of source rocks that had not yet been pushed into the oil window. Of course, some day those shallow source rocks will likely be buried deeply enough. Oil companies can be patient, but they can't be patient on a time scale of 10 million years.

I repeat that I do not have any confidential information about the Oregon-Washington drilling. However, if the oil companies saw either no source rocks or only immature source rocks, then even a global oil shortage will not bring them back. The Texas county with 30 dry holes followed by a big success is not the right model.

CHAPTER 3

Oil Reservoirs and Oil Traps

In western Pennsylvania during the early years of oil exploration, a favorite strategy was drilling next to a cemetery. It wasn't as ghoulish as it sounds. Cemeteries in western Pennsylvania were usually located on prominent hilltops. Sometimes the hilltops were there because sedimentary beds resistant to weathering had been "domed up": the surface hill could be the expression of an underlying structural dome.

The connection between a structural dome and an oil field was first appreciated in 1880.[1] Oil is less dense than water. Oil floats above water. If oil is present in the pores inside a sedimentary rock, the oil will tend to move upward and displace water downward. If the oil finds its way to a dome, a place where the rock layers are concave downward, the oil will be trapped in the dome. Most structural domes in the subsurface are not circular; they tend to be long and narrow. These elongated structures were named "anticlines" because the rocks dipped away from one another. (The reverse of an anticline is a syncline, a concave-up structure in which the rocks dip toward one another. Synclines are not a favorite place to drill for oil.) The concept of subsurface oil floating above water in a domed structure became known as the "anticlinal theory of oil." It was the first intellectual idea behind oil exploration.

In arid regions, in the desert, rock outcrops are plentiful. Out West, place names like "Circle Ridge" and "Racetrack Valley" were hints of anticlines. Exploration traditionally involved geologic mapping on the ground. When aerial photographs became available around 1945, geologists found that they could use the photos to map structure. (One lesson on the side: don't disturb the desert surface unnecessarily. Wagon roads abandoned more than 100 years ago are still visible on the aerial photos.) Usually, after a promising anticline turned up on the aerial photographs, an examination on the ground was necessary. I once found a dim, nearly straight line on an aerial photo that I couldn't explain. I wondered if it could be the surface expression of a fault line. Nope. On the ground, I found that one end of the line was at the end of a small hill; at the other end was a barn. As soon as the cows could see around the hill, they made a trail straight for the barn.

In the United States, by 1950 all the anticlines visible from the surface geology had been drilled. Methods other than surface geologic mapping were already in widespread use. Drilling the last surface anticline was not newsworthy. In contrast, in Iran today *all* the oil fields are surface-visible anticlines. Exploration in Iran and Iraq has been frozen at the same stage that the western United States reached in 1950.

I inadvertently turned up an example of the progression from drilling surface anticlines to finding subtle oil traps. Soon after the Nixon administration opened relationships with mainland China, a delegation of Chinese scientists (including some petroleum geologists) came to tour U.S. universities. Their first stop was the White House, and for purely geographic convenience, the second stop was Princeton. I got tagged to host the petroleum geologists. Since we were the first university on the list, we had only one clue to guide us. We were told not to ask the Chinese about equipment, because at the time China was woefully behind in scientific equipment. To avoid discussing equipment, we decided to talk about the intellectual basis underlying oil exploration. I was able to round up four Princeton geology professors with professional experience in the oil fields. (Today, sadly, that count is down to one.) The Chinese were familiar with, and were using, the latest ideas. The head of the Chinese delegation spoke

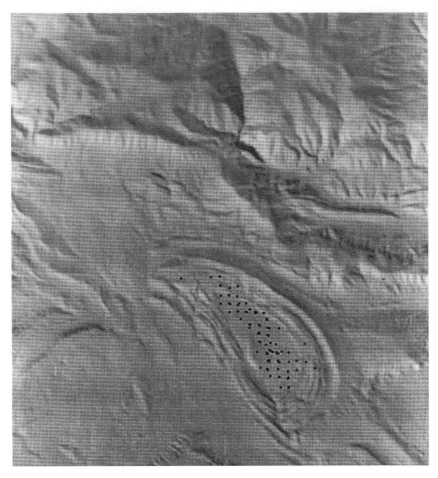

This land feature in central Wyoming was named "Circle Ridge" before the oil field was discovered. The oil wells are shown by solid black dots. The Circle Ridge oil field, discovered in 1917, has produced 30 million barrels of oil and is still producing half a million barrels per year.

disparagingly of the old-time Chinese geologists who could find oil only in the big anticlines in western China. Their activity had already shifted to subtle oil traps in eastern China.

For two days, I was busy hosting the Chinese. The next morning I was soaking in the bathtub when this thought hit me: oil production in China is going to peak earlier than in Iraq and Iran. China was further along in the exploration cycle. Hmmm . . . the CIA should be interested in that. I got dressed and walked to the office; the phone rang; it was the CIA. It had been shrewd enough to wait until the Chinese were gone before approaching me; I would not have agreed in advance to spy. Dealing with the CIA made me nervous. I decided to write out my observations instead of answering questions. (For one thing, the head of the Chinese petroleum delegation was from Szechwan. He loved Tabasco sauce. I didn't want the CIA sending him a case of poisoned Tabasco.) The Chinese were asking to import fast computers to process oil data. Fast computers can also run an air defense system or design nuclear weapons. The strategic question was whether the Chinese at that time had the expertise to use the computer output for oil exploration. The Princeton geologists voted resoundingly, "Yes!" In fact, we agreed that we had all four worked for worse exploration managers in American oil companies.

In 1953, the concept that oil would float above the water received a modification from M. King Hubbert. Hubbert showed that the oil-water contact would be tilted if the water were moving.[2] In the Rocky Mountain province, water is typically flowing through subsurface rock layers from the mountains toward the basins. Some Rocky Mountain structural "domes" weren't quite domes; one end of an anticline didn't plunge down to complete the trap. If the flowing water tilted the oil-water contact in the right direction, a trap could exist that would not be expected if the oil-water contact was exactly horizontal.

Because Hubbert worked at the Shell research lab, Shell had a head start in reevaluating the Rocky Mountain province. At that time, I had recently been hired by Shell, and I was pulled off its usual training program and sent to Wyoming to make accurate maps of several anticlines. The appropriate surveying equipment was called a plane table, which dated from the nineteenth century. The generation of geologists before me could, and frequently did, tell war stories about their years using the plane table. The overall Shell effort did some good, but it was not a major success for an odd reason. Fresh water flowing into sedimentary layers from the mountains contains the normal amount of dissolved oxygen. Where the fresh water encountered

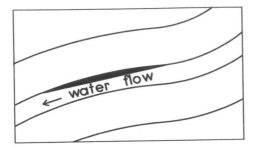

The contact between heavier water and lighter oil is ordinarily horizontal. When the water is actively flowing, the oil-water contact becomes tilted. Some structures that would not normally retain oil become oil traps if the water is moving.

oil, bacteria earned their living by oxidizing the oil. Provinces with heavy flows of fresh water into the subsurface gradually lose their oil to bacterial oxidation. However, I found it exciting to be applying an intellectually novel exploration idea two weeks after I received my undergraduate degree. I claim to have been the last petroleum geologist to use a plane table for a major oil company.

Active petroleum geologists spend 95 percent of their time looking for petroleum traps. Arguments about source rocks and thermal histories come up during coffee breaks; the core of the business is finding traps. If the traps are anticlines, today they are anticlines deeply buried beneath less folded rock layers. Some of them are not anticlines at all.

The East Texas oil field, discovered in 1930, caught everyone's attention. A well, drilled on a nongeologic hunch by old Dad Joiner, brought in the largest oil field in the lower 48 states.[3] East Texas was not an ordinary anticlinal trap. It turned out to be an "angular unconformity." Angular unconformities had been known since the very birth of modern geology. In the late eighteenth century, James Hutton described three angular unconformities in Scotland.[4] His interpretation, which geologists still accept, stated that older sedimentary layers had been tilted during mountain building, their upturned edges were eroded completely down to sea level, and then they were covered over by much younger layers of sediment. Hutton's great insight was that ordinary processes, going on today, could explain an angular unconformity, but only if geologic time extended over hundreds of millions of years. All three of Hutton's Scottish examples were smaller

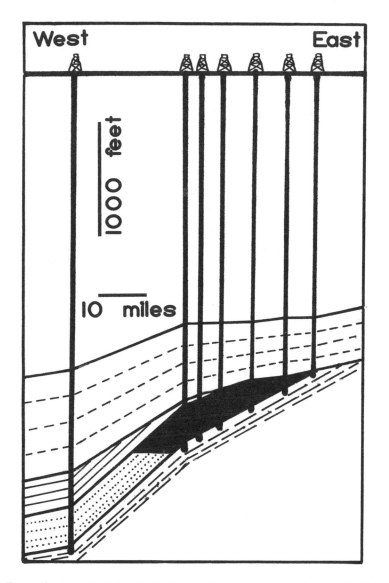

West
East

1000 feet

10 miles

East Texas, the largest oil field in the lower 48 states, is located where the Wood-bine Sandstone was tilted up, partially eroded, and buried by younger sediments. When East Texas was discovered by accident in 1930, this type of trap was not on the agenda of oil geologists. This diagram is an east-west cross section through the East Texas oil field.

than a football field. He would have been surprised, but delighted, to learn that an angular unconformity extended across four Texas counties and produced 5 billion barrels of oil.

Faults, which offset geologic layers, can be either good news or bad news. The good news is that faults can create a trap geometry by cutting off the upper ends of sedimentary layers. The bad news is that some faults allow oil and gas to leak up through the fault zone.[5]

Leaky and nonleaky faults became a major issue during the oil shortage of the 1970s and early 1980s. To encourage oil development in the United States, without giving huge price rewards for existing production, the federal government established two levels of oil prices. Oil from known oil fields was assigned a regulated price, but new oil could be sold at the much higher world price. The initial regulations were very poorly written. New oil was defined as coming from a different formation. Unfortunately, bureaucrats and geologists used different definitions for *formation*. The Department of Energy definition was the same as "We went to Yellowstone and saw the geologic formations." Over in the Interior Department, the U.S. Geological Survey had codified *formation* to mean a named, mappable, unit of sedimentary rocks:[6] the Mowry Formation, the Phosphoria Formation. Oil companies pounced on the USGS meaning and gleefully reported any oil produced from an additional producing formation, even within an existing oil field, as "new" oil. Even after the initial regulations were rewritten, controversy continued. The Department of Energy eventually filed lawsuits claiming that several major oil companies had misrepresented existing oil as new production.

Through a university friend, I was asked to act as an expert witness for the federal government. Given my ancestral roots in the oil business, I was a bit reluctant. As a university professor, I had no legal or ethical restrictions to keep me from testifying against a major oil company. The scales were tipped when I learned that the defendant was Texaco. I had competed against Texaco practically all my life; there was no reason to stop now.

Texaco's evidence for "new" oil was dividing the oil trapped around a single Gulf Coast salt dome into more than a 100 indepen-

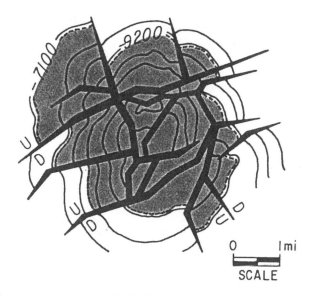

SCALE

As a salt dome moves upward, faults form in the surrounding sedimentary layers. Oil trapped against the dome has to be located separately in each of the
blocks between the faults. This is a map view of the Anahuac salt dome.

dent oil fields. Salt domes, on the Gulf Coast and elsewhere, arise
because deeply buried beds of sodium chloride—ordinary table salt—
become softened with temperature and rise up toward the surface.
(The best-known salt dome is Avery Island, Louisiana. It is at once an
oil field, a salt mine, an island that rises a few feet above the swamp,
and a great place to grow hot peppers. Check the label on your Tabasco
bottle.) The salt rising toward the surface creates cracks in the surrounding sediment layers, rather like cracks around a bullet hole in a
glass window. Texaco's claim was that every crack-bounded fragment
was an independent oil field. There were two questions about the Texaco position: (1) Was one salt dome a unified entity like one bullet
hole, and (2) Were the cracks sealed or leaky?

The evidence Texaco supplied to the court was in two parts. A
former president of the American Association of Petroleum Geologists
signed an affidavit listing more than 100 independent oil fields in
each Texaco salt dome. The second part was a room full of boxes con-

taining individual reservoir descriptions in the numerical order in which they had been filed with the state of Louisiana regulatory agency. There were five salt domes and more than 100 claimed oil fields in each salt dome; it took two of us two full days just to index the papers. After the federal lawyers filed initial replies, the Texaco lawyers wanted to compare the "credentials" of the opposing experts. They claimed that I was merely a "sedimentary petrologist," whereas their man was a recognized petroleum geologist. Ever after, whenever I taught sedimentology, I told my students that I was a court-certified sedimentary petrologist.

Using my homemade index to the Texaco filings, I could match up maps of adjacent "independent" oil fields. Almost all the maps included the elevation of the oil-water contact, known to an accuracy of about one foot. Where two portions of the same oil-bearing sandstone were separated by a fault, the oil-water contact elevations were sometimes identical across the dividing fault, identical to the exact same foot. It meant that the adjacent reservoirs were connected. A pressure difference of one pound per square inch across the fault would offset the oil-water contacts by 20 feet, but the offset was less than a foot. (Air pressure at sea level is 14.7 pounds per square inch.) Also, there were over-and-under pairs of reservoirs with identical oil-water contact elevations. Some of the faults were demonstrably leaky.

My affidavit pointed out these identical oil-water contacts as evidence that the "independent" oil fields were physically connected. Another geologist, also working for the federals, compiled a much longer list of connected reservoirs. The Texaco lawyers replied with additional experts and claimed that their maps were never intended for the purpose for which we used them. I thought that Texaco made a pretty good reply, but one of the federal lawyers gave me her take: "When you have to withdraw the evidence on which your expert based his opinion, your goose is cooked." Apparently, the height of your expert's standing does not matter when the rug is pulled out from under him or her. I heard nothing more about the case while it went through the inevitable appeals. Then one morning my copy of the *New York Times* carried the story, front page, above the fold; Texaco was fined more than a billion dollars.[7] My fee bought me a bottom-of-the-line Volvo.

The most unusual oil fields are meteorite scars, one in North Dakota, one in Oklahoma, and possibly others. Initially, they were puzzles, but further analysis showed them to be scars of meteorite strikes buried beneath younger sedimentary rocks. The shattering effect of the meteorite impact opened up abundant cracks in the rock, which later filled with oil.

Anticlines, angular unconformities, salt domes, and fault traps are all structural features. There are also oil fields unrelated to structure. When any of us look at sedimentary strata in a road cut, we get the illusion that the layers go on forever. They don't. Sediments are deposited by rivers, ocean currents, and organic growth. A sediment layer has to begin and end somewhere. Beginnings and endings of layers can trap oil. Patterns of strata are called "stratigraphy"; oil traps formed by strata are called "stratigraphic traps," the usual abbreviation being "strat trap."

The most spectacular strat traps are reefs. A "reef" in this sense includes modern coral reefs, although the dominant corals in today's coral reefs arose only during the past 20 million years. Rock patterns that resemble modern reefs go back 500 million years, but there are continuing arguments about whether these were all "reefs" in the sense that organisms grew a rigid framework to form the reef. One of my colleagues at Shell convinced me that the great Capitan Reef in west Texas (which includes Carlsbad Cavern) was in fact the great Capitan mud mound.[8]

There are ancient reefs, some of them oil fields, that probably fit the organic-framework definition. Oil production rates from reef rock can be spectacular. The well traditionally listed as the largest in history, flowing more than 100,000 barrels per day, was in the Golden Lane, a buried reef in Mexico. (A nice footnote: the location for the big well was chosen by Everett Lee DeGolyer while he was still an undergraduate at the University of Oklahoma. It was a fabulous start to DeGolyer's long, distinguished career.) Reefs of all sizes exist. Leduc and Redwater beneath the plains of Alberta are huge fields; midsize

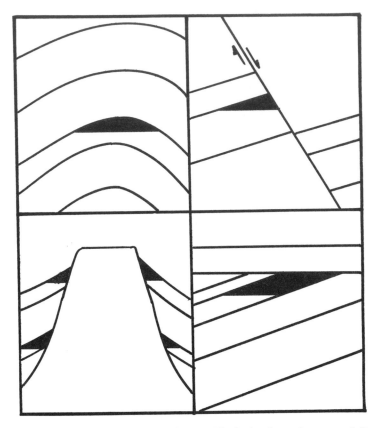

Structural oil traps take a variety of forms. Clockwise from the upper left are an anticline, a fault trap, an angular unconformity, and a salt dome. Oil accumulations are shown in solid black.

reefs were found to be highly productive in Scurry County, Texas, and small patch reefs produce oil in Michigan.[9]

Stratigraphic traps are also found in sandstones. The first to be appreciated were the long narrow "shoestring sands" beneath northeastern Oklahoma and southeastern Kansas. Several geologists in the 1920s recognized the analogy between the shoestring sands and the offshore bars along the Texas and Atlantic coasts, as at Galveston and Atlantic City.[10] A second type of sand geometry was eventually associated with the "point bars" deposited on the concave sides of river meanders.[11]

This core, from a well in the Leduc reef in Alberta, shows excellent porosity. This is the rock that oil geologists see in their sweetest daydreams. Replacement of the original limestone by dolomite has obscured most of the original texture.

Finding stratigraphic traps is not easy. Sometimes, but rarely, there are visible surface hints because the sediments above are subtly draped over a subsurface reef or a sandstone body. We joke about the princess who could feel a pea concealed beneath 20 feather mattresses. Other means of locating strat traps are explained in a later chapter. For now, suffice it to say: the very last of the world's oil will be discovered in stratigraphic traps.

A structural or stratigraphic arrangement that ought to trap oil is of no use if the drill gets to the target and the rock is hard, tough, tight, and devoid of porosity. Old-timers called it "suitcase rock": time to pack up the suitcase and go home. Today's geologist may get the data over a satellite link to the 20th floor of a Houston office tower, but the news is still bad. Oil does not come from big caverns underground; it comes from rock, and the reservoir rock has to contain and deliver oil.

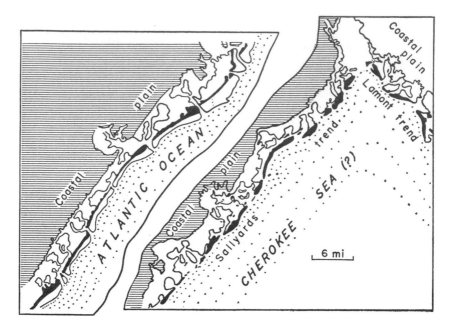

In map view, the shoestring sands of eastern Kansas and Oklahoma, 300 million years old, strongly resemble the present-day offshore sandbars of the Atlantic and Gulf coasts.

"Contain" and "deliver" emphasize the two dominant characteristics of a reservoir rock. "Contain" refers to porosity, the fraction of the rock that is open pores. "Deliver" depends on the size of connections between the pores. It's like you just won the big lottery; $10 million, but they pay it out over 20 years. $10 million is the porosity; 20 years is determined by the permeability.

Porosity and permeability do not always occur together. Gas bubbles in a solidified volcanic lava flow have porosity but no permeability; the bubbles are not connected to one another. The first oil wells in Nevada produced from volcanic rock, but it was not a conventional lava flow.[12] Glowing avalanches that come down the sides of volcanoes can be so hot that the grains weld solidly to one another, forming a "welded tuff." A glowing avalanche wiped out the town of St. Pierre on the Caribbean island of Martinique in 1906: 30,000 fatalities, 4 survivors. However, the St. Pierre glowing avalanche was not

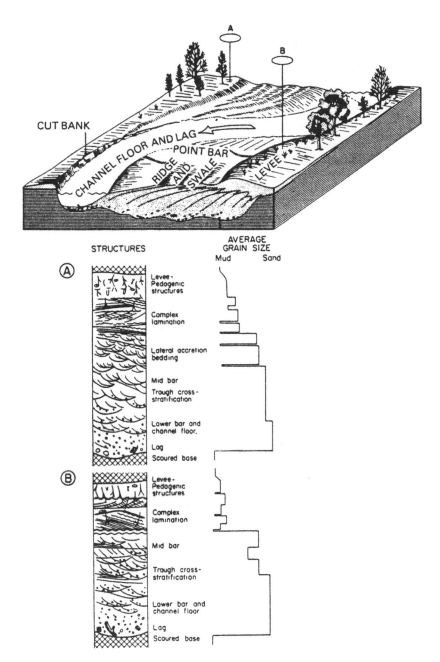

During the maximum stage of a major flood, a river picks up large amounts of sediment. As the flood wanes, a distinctive sequence of sediments is formed. Pebbles are dropped out first and form the bottom of the sequence. Above are sands with scoop-shaped layers, followed by sand with ripple marks, sand with parallel layers, followed by mud. The sediments gradually get finer grained as you go upward through the sequence.

hot enough to weld. If it had formed a welded tuff, there would have been 30,004 fatalities. The relevance to the oil business is that the pores in a welded tuff are connected to one another. And not just to the oil business: the Round Mountain gold mine in Nevada initially contained 16 million ounces of gold, deposited by hot water flowing through a welded tuff.[13]

The concept of permeability, along with the law of fluid flow through a porous solid, was developed in 1850 by Henri Darcy, in Dijon. The flow law is known as Darcy's law, and the unit of permeability is the darcy. (Scientific immortality comes from being shifted to lowercase: watt, ohm, ampere. An einstein is a mole of photons.) The darcy turned out to be a rather large unit; the millidarcy, 1/1,000 of a darcy, is universally used in the oil fields. (A millihelen is a face that would launch one ship.) Typical oil field language: "We got 80 feet of saturation and two hunnert millidarcies. We're cuttin' up a fat hawg."

When I stopped with my family in Dijon, everyone but me wanted to buy those little pots of mustard to take home to our friends and relations. I was delighted that the city square was "Place Darcy" and that the square contained a two-story monument to Henri Darcy. On the memorial was no mention of Darcy's law, no plaque from petroleum geologists celebrating the millidarcy. I knew why. Darcy was the city water engineer of Dijon. In 1850, Dijon was the culinary capital of the world. But the chefs were ripping off their tall white toques in anguish because dirty city water was making their aspic cloudy. In order to design sand filters for Dijon, Darcy first did some small-scale experiments. The successful sand filters won Darcy a two-story monument in the city square. The equation from his experimental results won Darcy lowercase immortality.[14]

Petroleum reservoir rocks fall into two main categories: sandstones and carbonates. Carbonates get their name because they are made up of either calcium carbonate (limestone) or calcium-magnesium carbonate (dolomite). There were numerous coffee-break arguments among the petroleum geologists at Shell about whether sandstones or carbonates contained more oil. Finally, a student summer hire was sentenced to spend an entire summer in the Shell files to pick the winner. Sandstones won in North America, but with the Middle East included, carbonates came out ahead.

In one sense, sand deposits are familiar. Almost everyone has seen a river and has gone to the beach. A river-deposited sand has an interesting internal anatomy; most of the sand is deposited in the waning phase of a river flood. At flood stage, a river carries an enormous amount of sediment; as the flood slows down, the larger grains settle out first. Any pebbles that are present form the bottom layer. Coarse sand grains are deposited next in distinctive scoop-shaped patterns. Finer sands follow with wavy layers and parallel layers; mud is at the top. This whole package, typically deposited on the inside of river bends, is known as a "point bar" sequence.[15] The name goes back to the old Mississippi River captains: "I know every sand bar on this river." WHOMP. "There's one of them now."

At the beach, an "offshore bar" is a mix of beach sand, wind-blown dune sand, and sand deposited underwater. Offshore bars have their largest sand grains at the top, in contrast to river point bars, with pebbles at the bottom. The distinction between offshore bars and point bars is put to practical use in the oil patch. An offshore bar runs parallel to the ancient shoreline; the rivers that deposit point bars are flowing roughly perpendicular to the fossil shoreline. Following trends (cheerfully called "trendology") is a favorite sport, but the direction of the trend depends on whether you are chasing a point bar or an offshore bar.

Instead of being confined to bars, some ancient sandstones extend for hundreds of miles in all directions. The obvious first guess is that "blanket sandstones" are the dunes from ancient deserts, a guess that may well be correct. Unlike what you see in old movies, most deserts are not composed of sand dunes. Wind blows the sand and dust away, leaving behind gravel, bare rock, and some scruffy vegetation. Of course, the sand does have to end up somewhere. The Empty Quarter of the Arabian Peninsula, the Rub' al Khali, is a sand sea that could be the model for blanket sand deposits.[16]

Ancient dune sandstones are deposited mostly on the steep downwind side of the dune, where sand is accumulating. These inclined sand layers are known as "crossbeds" because they are tilted crossways to the horizontal beds in adjacent layers. I got a real thrill from crossbeds on the north coast of Arran, in Scotland. The beds were 270 million years old, from a time period named "Permian." My guidebook showed that Arran was about 10 degrees north of the equator during Permian time.[17] That told me which way the trade winds would have been blowing; the crossbeds were dipping exactly in the ancient downwind direction. I ignored the cold spitting rain and walked across those ancient dune sands feeling the warm Permian trade wind at my back.

Before 1950, almost all geologists expected that no sands would make their way into deep water. From 1950 to 1965, that expectation changed with the discovery of sand transported by turbidity currents.[18] Turbidity currents were mud-sand-water suspensions that flowed downhill beneath the water surface. Stirring mud or sand into the water increased the density, and the denser suspension flowed underwater, attaining speeds on steep bottom slopes of 40 miles an hour. When the flow reached a flatter bottom, the flow slowed down, and sand and then mud settled out. The arrangement with the finer grains on top is similar to river point bar sands, but there is a long list of details for distinguishing point bar deposits from turbidity current deposits.

Turbidity currents are usually attributed to the buildup of sediment in shallow water, with the sediment stirred up either by a slump or by an earthquake. An unresolved issue is whether the sediment load from a large river at flood stage, or from a flash flood on a smaller river, could be injected directly beneath the sea. Sandstones deposited by turbidity currents, called turbidites, are not pretty. Usually some mud is mixed in with the sand, which lowers the permeability. The turbidity current itself flows around obstacles, making individual sands difficult to follow. Nonetheless, the Ventura Basin just north of Los Angeles produces major amounts of oil from turbidity current sandstones.[19]

Although sand is a familiar sediment, carbonate sediments are deposited only in the tropics and subtropics. This has a delightful by-product. When I taught sedimentation, it was intellectually necessary for my class to adjourn to the Bahamas or the Florida Keys for spring break. The first time I went on a one-week field trip to see carbonate sedimentation, I learned more that week than any week in my life. Bob Ginzburg took a group of Shell geologists to places where carbonate sediments were actively being deposited, sediments that could later turn into limestones. At the end of the week, Ginzburg took us snorkeling on a coral reef. The reef was fabulous: different kinds of corals, sea urchins, gloriously colored fish. Slowly the message sank in: all those textbook pictures and museum reconstructions of ancient environments were heavily biased toward reeflike environments. Most limestones were deposited in less spectacular places like Florida Bay. In his Socratic, University of Chicago style, Ginzburg never stated the conclusion outright; he simply arranged the trip to let the sediments preach the word. Later, I shamelessly plagiarized Bob's course content for my own students.

The dominant modern carbonate sediment is calcium carbonate. Fine-grained calcium carbonate mud usually gets consolidated into massive limestones, usually with little or no porosity—suitcase rock. The most massive cliffs in the Grand Canyon, about halfway down, are formed by the Redwall Limestone, deposited originally as a calcium carbonate mud.

About 10 percent of ancient limestones do have porosity. Carbonate sands are usually either broken fragments of seashells or sand-size grains that look like fish eggs. They aren't fish eggs; they are sand-size grains that grow today where tides bring water on and off the edges of the Bahama Banks. The resemblance to fish eggs gave these sediments a name, oolite, from the Greek for "egg." Oolite is pronounced with the two *o*'s as two different syllables; see your dictionary.

Most massive and nonporous limestones contain textures made by invertebrate animals that ingest sediment and turn out fecal pellets. Usually, the pellets get squished into the mud. Rarely do the fecal pellets themselves form a porous sedimentary rock. In the 1970s, the first native-born Saudi to earn a doctorate in petroleum geology arrived for a year of work at Princeton. I used the occasion to twist

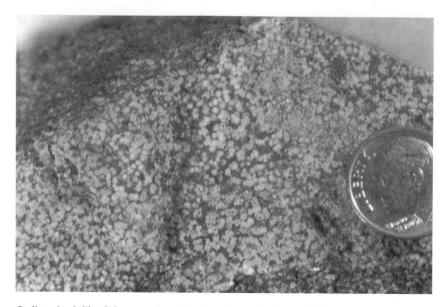

Oolites look like fish eggs, but they are formed where tidal currents move sediment at the edges of shallow-water banks. If the porosity is preserved, an oolite can be a productive oil reservoir rock.

Aramco's collective arm for samples from the supergiant Ghawar field. As soon as the samples were ready, I made an appointment with our Saudi visitor to examine the samples together using petrographic microscopes. That morning, I was really excited. Examining the reservoir rock of the world's biggest oil field was for me a thrill bigger than climbing Mount Everest. A small part of the reservoir was dolomite, but most of it turned out to be a fecal-pellet limestone.[20] I had to go home that evening and explain to my family that the reservoir rock in the world's biggest oil field was made of shit.

The dominant carbonate reservoir rock is not calcium carbonate; it is a calcium-magnesium carbonate named dolomite. The spectacular Dolomite Alps did not give their name to dolomite; the mountains were named for the mineral. Dolomite was named for Count Deodat de Dolomieu, a Frenchman who died in an Italian prison in 1801.

Dolomite was a long-standing problem. When I was a freshman, textbooks stated that dolomite was not forming today.[21] Almost everything about geology can be understood by extending processes that are active today to the geologic past. There was a very short list of rocks that did not seem to be forming right now: oil source rocks, glowing ash flows hot enough to make welded tuffs, and dolomite. In rocks older than 350 million years, more than half of the carbonate rocks were dolomite instead of limestone. It was as if something would zap limestones and convert them to dolomite. The older the rock, the more likely it was to have been bitten by the "dolomite bug."

When I started work at the Shell research lab in 1958, I was told that I ought to work on dolomite. The Shell managers explained that almost no progress had been made since F. M. Van Tuyl's publications in 1916. Small world! Professor Van Tuyl taught me freshman geology at the Colorado School of Mines. In 1916, he identified two types of dolomite.[22] One was fine grained, contained no fossils, and was found in layers parallel to the other sedimentary layers. The second was coarse grained, fossils existed usually as hollow molds where the seashells used to be, and the dolomite formed irregular masses that cut across the sedimentary layering. There it stood for 40 years, with no way to study the problem in modern sediments.

The Central Basin Platform, an area of 4,000 square miles beneath the plains of west Texas and New Mexico, is a dolomite mass 6,000 feet thick.[23] Drill crews called it "the mile of dolomite." It was tough rock to drill; the drillers hated it. Geologists loved it. Around the margins of the Central Basin Platform were some of the richest oil fields in the United States, a "little Middle East." Understanding dolomite in general and the Central Basin Platform in particular was an important research target.

My colleague F. Jerry Lucia made the first breakthrough. He noticed that some of the subsurface dolomite from the Central Basin Platform had a tiny amount of natural radioactivity that he could not explain. Jerry coated a thin slice of the rock with a photographic emulsion, stored the coated rock for a month in the dark, and then developed the photograph. Usually the tracks left by radioactive decay in rocks come from specific grains like zircons; under the microscope the radioactive decay tracks in the emulsion make a small zircon grain

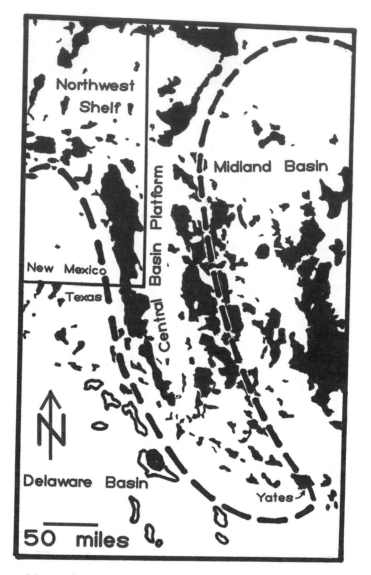

The Central Basin Platform, in west Texas and southeast New Mexico, is a 250-million-year-old analogue of the modern Bahama Banks. Major oil fields tend to be around the margins of the Central Basin Platform. The Yates field had a number of spectacularly productive wells.

look like a tiny porcupine. Lucia found the opposite: the decay tracks were coming one by one out of the dolomite, and he guessed that a small amount of uranium had been incorporated into the dolomite. He then wanted to know if the decay of uranium to lead would give him a clock to learn when the dolomite had been deposited. At that time, there were only three or four labs in the United States equipped for measuring uranium-lead dates. At the Shell lab, one of them was down the hall.

The uranium date on the dolomite came out to be 275 million years. The time when the Central Basin Platform sediments were deposits was essentially the same. This dolomite had to be formed pretty soon (within a million or two years) after the sediments were deposited. It was not a case of "zapping" a much older limestone to make dolomite. I tried some calculations. Converting a calcium carbonate (limestone) to calcium-magnesium carbonate (dolomite) clearly required water with plenty of magnesium. The most magnesium-rich water known was the brine in the Dead Sea. I did a calculation using Darcy's law. How long it would take for Dead Sea brine to convert the Central Basin Platform from limestone to dolomite? Fifteen million years. Lucia's age measurement said that 15 million was too long. Converting the Central Basin Platform to dolomite must have been a pay-as-you-go process, with dolomite being formed right along as the sediments were deposited.

A bit of good luck intervened. Ray Murray of the Shell research staff gave a talk at Princeton as part of the corporate recruiting effort. He gave a guest lecture that included Van Tuyl's 1916 calculation that all dolomite rocks should start out with at least 12 percent porosity. A Princeton professor, Alfred Fischer, objected, saying that he had a specimen of a very young dolomite from Curaçao with less than 12 percent porosity. A few weeks later I visited Princeton and talked Fischer into letting me have half of his Curaçao specimen. Fischer said that he had also seen some interesting things on Bonaire, an island adjacent to Curaçao. Back in Houston we made four microphotographs: two from the young Curaçao dolomite and two from the old Central Basin Platform. We circulated the photos around, challenging people to pick out which was which. The microphotographs were essentially identical; Murray was the only one who could tell them apart. (Crystal edges on the young dolomite were a tiny bit sharper.) On the strength of the photographs and the constraint of the uranium date, Shell management agreed to send Jerry Lucia, Peter Weyl, and me to Curaçao and Bonaire for three weeks.

Curaçao and Bonaire are overseas portions of the Netherlands, like Hawaii is now one of the United States. Although they are tropical islands, they were as neat and clean as a Dutch village. At the time, 1961, Shell on Curaçao was processing Venezuelan crude oil in the world's second-largest oil refinery. Lucia and I started work on Curaçao; a week later we were joined by Weyl, and we moved on to Bonaire. At the research lab, Weyl described himself as a physicist in the chemistry department working on geology. Peter had been a refugee from Nazi Germany and was educated in the "golden age" of the University of Chicago after World War II. I learned more from Peter than any scientist I have ever worked with.

Bonaire was a geologist's dream. On the south end of the island was a salt lake and salt marsh, made salty by the evaporation of seawater. White chalky crusts from south Bonaire turned out to be dolomite. Back in Houston, the lab ran carbon-14 ages on the carbonate carbon in the dolomite, and the ages were 1,400 and 2,000 years. The first marine-related modern dolomite! My freshman textbook was wrong.

Almost nothing could live on the salt flats; the water could be 10 times saltier than seawater, and in a few minutes a rain shower could take the salinity to zero. The dolomite crusts ran parallel to the flat surface, only microscopic organisms could survive there, and the dolomite crystals in the crusts were very small. South Bonaire was producing Van Tuyl's first type of dolomite.

The north end of Bonaire was made of older rocks, but still young on the geologic time scale: less than a million years old. On north Bonaire were coarse-grained dolomites, with hollow fossil casts, in large-scale blobs that cut across the layering in the surrounding limestone: Van Tuyl's second style of dolomite. It took only a tiny amount

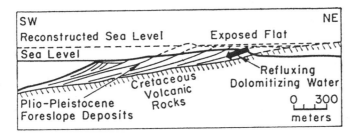

Insiders often refer to a conceptual drawing as a "cartoon." This cartoon suggests that an earlier version of present-day south Bonaire existed above the rocks now exposed on the north end of Bonaire. Reprinted with permission from *Science* **143**:679; copyright © 1963 AAAS.

of geologic imagination to infer that the process that made dolomite on north Bonaire was probably occurring today beneath south Bonaire. Those surface salt brines are denser than seawater, and the brines probably drain downward to convert part of underlying limestones into large blobs of dolomite. Put the other way, in the recent past, there was a south Bonaire running atop the present-day north Bonaire.[24]

Bonaire had a third surprise. Van Tuyl had calculated that a freshly formed dolomite should have at least 12 percent porosity, the conclusion that Al Fischer had disputed. We did find some dolomite from north Bonaire, inferred to be young, with porosity too low to qualify as an oil reservoir. I won't forget collecting the lowest-porosity sample. It was tough outcrop, I was swinging my rock hammer with all my strength, and I made the mistake that geology students are warned about. A rock hammer has a blunt end for breaking rock and a pointed end for prying. I caught the hammer on an overhead branch, and the pointy end came down hard on top of my head. I'm here to tell you that it doesn't cause instant death, but scalp wounds bleed like crazy.

After a three-year delay required by the Shell lab, we published papers describing the two major types of dolomite on Bonaire. What we did not publish was the bit about low porosity, which was the most useful part to Shell. Van Tuyl's porosity calculation assumed that magnesium was swapped for calcium, but no carbonate entered or left the rock. The low-porosity rock, from a blood-soaked sample bag, showed that salt brines do introduce a small amount of carbonate and that rocks immediately under the salt flats are most likely to emerge with low porosity: porosity too low to make a decent oil reservoir. This explained why the best oil fields were around the *edge* of the Central Basin Platform. Most of the dolomite in the center had low porosity. A useful rule of thumb emerged. If you find low-porosity dolomite, move out toward the edge of the carbonate platform. If you find limestone, move in closer to the platform edge.

After that three-week trip, I never went back to Bonaire. The fabulous natural laboratory on the south end of the island was bulldozed to make commercial salt evaporation pans. Luxury hotels were built to turn Bonaire into a scuba diver's paradise. It is no longer a geologist's paradise.

A reservoir rock with both porosity and permeability is a necessary ingredient for a successful oil field. Another necessary ingredient is the opposite kind of rock, one that keeps the oil and gas from simply migrating on up to the surface. The sealing layer above the oil reservoir is known in the trade as a "cap rock." A common mistake is to tag the first rock above the reservoir rock as the cap rock. Often, the rock immediately above the commercial reservoir rock is simply a noncommercial reservoir rock: a rock that will not produce oil fast enough to make money but one that would leak out the oil and gas over geologic time. The real cap rock may be farther on up the column.

There are two ways that a rock can function as a cap rock. If the holes in the rock are very tiny, the oil cannot push its way into the pores because of the surface tension at the oil-water interface.[25] In mudstones, also called shales, holes are usually 1 micron across (1/1,000 of a millimeter, 40 millionths of an inch). An easy calculation compares the oil-water surface tension around a 1-micron hole with the pressure due to the oil column. The calculation shows that, at best, we can expect shale to hold back the buoyancy of a column of oil 200 feet high. When the oil column is taller than 200 feet, we begin to expect that something more effective than mudstone is retaining the oil. One of Shell's discoveries in the North Sea had more than 200 feet of oil-saturated rock. (You might expect that oil companies would know something about managing secrecy. Bids were due soon on tracts adjacent to the discovery. Shell plugged and abandoned the discovery well and pretended it was a dry hole. Before the bids went in, only three people in the entire oil company knew it was a major discovery.)

Super oil columns like those in the North Sea and the Middle East require a super category of cap rock. The world champions among cap rocks are two related rock types: anhydrite and salt. Both are formed from the evaporation of seawater. Both start out with porosity, but before they are buried to the top of the oil window, each converts to a rock with essentially zero porosity and zero permeability.

Anhydrite starts out as the mineral gypsum. Gypsum consists of calcium sulfate and includes two water molecules for each molecule of calcium sulfate. Gypsum layers, formed when seawater is concentrated by evaporation, lose the two water molecules at a depth of about 1,000 feet, and the layer converts to a hard dense layer of anhydrite. ("Anhydrite" contains no water; it is *anhydrous*. Too bad all mineral names aren't so easy to remember.)

Further evaporation of seawater, after the gypsum comes out, precipitates ordinary table salt, sodium chloride. Both modern seawater evaporation ponds and ancient salt deposits supply us with salt. (Of course, the salt industry has ancient roots. Our word *salary* comes from "a man worth his salt." Salzburg, Mozart's home city, took its name and its wealth from the salt mines.) For an interesting reason, the world's deepest salt mines are only 3,000 feet deep. The deepest metal mines go to 12,000 feet. Deeper than 3,000 feet, salt starts to flow. A mining engineer told me about walking around in a 3,000-foot salt mine in Manitoba while the equipment was turned off and the mine was quiet. He could hear the salt creaking and groaning as the walls slowly tried to close in. Guaranteed claustrophobia attack.

As with anhydrite, closing the original pore space in salt happens at depths shallower than the top of the oil window. When oil starts to crack out of the source rocks, anhydrite and salt beds are tight seals. As you would expect, the Middle East oil fields come supplied with cap rocks of salt and anhydrite.

This fragment from an oil well core is made of the mineral anhydrite: calcium sulfate. The sediment was deposited originally as gypsum: calcium sulfate with two water molecules. At a depth of about 1,000 feet, the water molecules are lost, and the resulting anhydrite has zero porosity. The great oil fields of the Middle East have anhydrite layers that seal the oil and gas in porous rocks beneath.

There is a leakage test for cap rocks even more profound than retaining oil: helium gas. Helium sources are widespread, but helium gas fields are quite limited. Ordinary sedimentary rocks carry about one part per million of uranium. During the slow decay to lead, each uranium atom spits out six to eight alpha particles. An alpha particle in physics is identical to the nucleus of a helium atom in chemistry. The helium gas that we put in party balloons is simply used alpha particles. You have probably noticed that rubber party balloons don't float after a day or two. The helium leaks right through the rubber.

Helium is leaky stuff. Helium does not form ordinary chemical bonds, and helium atoms are tiny little devils. A standard test for subtle leaks in high-tech equipment uses helium. Helium produced by uranium decay is retained in the same traps as natural gas.[26] All our commercial helium is a by-product of natural gas production. Helium is the ultimate nonrenewable resource. The helium from party balloons eventually escapes out the top of the Earth's atmosphere into space.

Almost invariably, natural gas fields with recoverable helium have anhydrite or salt cap rocks. One of my oil company colleagues claimed that he knew of only one well with significant helium that was not capped by anhydrite or salt. So, friends, where would we go to store our radioactive wastes? Right! The military radioactive waste disposal site, WIPP, is located in a salt bed: helium leak-tested for geologic time.[27] The proposed Yucca Mountain disposal site has neither anhydrite nor salt.

There is a useful, but less drastic, type of leak test. As a natural gas field is depleted, there are two scenarios. Either the wells eventually go over to producing water, or the gas pressure may simply drop until no more gas is produced. In the second case, we have an empty and isolated subsurface reservoir. Gas pipeline companies use a few of the smaller depleted fields as underground storage tanks. For years I have been suggesting that my students write up a business plan for converting a depleted gas field into a deep disposal site for liquid chemical wastes. Waste disposal can be a profitable business; there are very few certified disposal wells for chemical wastes.

Despite our improved understanding of source rocks, the oil window, oil traps, reservoir rocks, and cap rocks, 9 out of 10 exploration wells are dry holes, sometimes called "dusters." Actually, most of them are not literally dry; they contain salt water. A "wet hole" is equally bad news. If we are so smart, why aren't we rich? If oil fields already exist 20 miles away in all directions, then the source rock–oil window question is already answered. But the success of exploration wells is bad even in certified oil country.

Murphy's Law: if anything can go wrong, it will. Deffeyes' Rule: if even one thing goes wrong, you get a dry hole. There are lots of ways to lose. A proposal for drilling an exploration well always contains a plausible scenario that (1) is consistent with all the known facts and (2) would contain an economically valuable amount of oil or natural gas. Writing and evaluating these scenarios is as intricate as a Balanchine ballet. Too pessimistic and you never drill an exploratory well. Too optimistic and the evaluator gets the giggles.

When I was young, I was shocked to hear that a respected manager had ordered one of his junior geologists to draw up maps containing a million-barrel oil field in the area around a recent discovery. The only restriction was that the maps should not misrepresent the known data. I felt that the junior geologist had been ordered to commit intellectual fraud. Later, I came to understand. After the manager had given the same assignment to several staff members and none of them had concocted a plausible million-barrel scenario, the manager sold the oil field.

There is a spectrum of things that can go wrong, from big conceptual errors to small detailed mistakes. All of them are fatal.

An example of a conceptual error involves timing. A structural trap results from folding and faulting at a particular time in the geologic past. Burying the source rocks into the oil window happened at a different time. If the trap formed *after* the oil was generated, tough luck. We don't always know exactly when these times were, but we always have to worry about it.

Here is an example of an error in a detail: for an investment program, I evaluated a proposal for a deep well in Louisiana. In my analysis, I mentioned that the fault labeled "B" on the map could possibly be a bit farther north. On balance, it looked like a good prospect, and I recommended that the investment program take a share in the well. When the well was drilled, it turned out that God had a way of locating fault "B" a whole lot farther north; we wound up on the down side of the fault. Dry hole.

All too often, the well is dry even when there is no obvious mistake. The structure in the well matches the predrilling scenario, the reservoir rock has porosity and permeability, but the well contains salt water. The surgery was a success, but the patient died. In my opinion, leaky cap rocks are the most common cause of these cryptic failures. Even with the best anhydrite or salt for a cap rock, one fault that does not seal can empty the reservoir.

It still isn't easy to find oil, but understanding source rocks, the oil window, traps, reservoir rocks, and cap rocks does help. It beats drilling next to cemeteries.

CHAPTER 4

Finding It

A lady goes into a fabric store and enlists the salesclerk's help in picking out a nice fabric for a nightgown. After they find something suitably frilly, the customer asks to buy 17 yards. Naturally, the salesclerk asks what kind of nightgown needs 17 yards of fabric. "Oh, my husband is an exploration geologist. For them, looking for it is as much fun as finding it."

Finding it influences the price of gasoline and (indirectly) the price of groceries. Economists and geologists start from completely different viewpoints. Economists state that oil discoveries depend on the level of investment put into the search.[1] Geologists claim that showing up at the cashier's cage with dollar bills does not increase the amount of oil in the ground. As usual, both are partly correct. For zero investment, you find zero new oil. The geologists' bias is based partly on intuition and partly on experience. Pouring money into the search in a hurry does get more wells drilled, but many of the added wells are "turkeys," locations with low probabilities for finding oil. Chapters 7 and 8 explain methods of estimating future oil supplies, independent of an economic or a geologic bias.

Present-day methods of finding oil differ a lot from the old-timer out mapping with a plane table. I mentioned in a previous chapter that oil geologists, early on, examined essentially all the land area on this planet. Here is another example: in 1978, I was watching while three Princeton geology professors discussed a possible research project in the Zagros Mountains of Iran. (Not long afterward, the Ayatollah Khomeini took over Iran, and the research plan was forgotten.) The three geologists had worked for three different oil companies; none was a specialist on the Middle East. As they went over the map, every valley they pointed to had been visited by one of the three. I was dazzled: collectively, they had examined every major valley in the Zagros.

The most difficult land areas to explore are those under heavy jungle with persistent cloud cover. Aerial photos usually do not exist; if they do, they don't show much. However, side-looking radar does an excellent job of revealing the terrain. Geologic structure is visible on side-looking radar images of the New Guinea jungles.

Surface exploration was largely complete, worldwide, 40 years ago. There are two major remaining exploration methods: subsurface geology and seismic exploration. These two methods dominate the industry today. Lots of other procedures have been suggested, and some of them are "snake oil": methods without a scientific basis. If I included a specific list of snake oils, libel lawyers would form a double line on my front walk.

Subsurface geology uses information from already drilled wells. The previous wells were drilled on surface anticlines, by drilling along trend lines defined by successful wells and by random dry holes. By long-standing custom, information from existing wells is made public. In some states, regulations require that the data be made public.[2] The underlying motivation is the hope that we all will find more oil if we all share the raw data. If a well is being drilled adjacent to property that is still on the market, the well can be declared a "tight hole" and the information release delayed for two years.

As a well is drilled, mud is circulated down the drill pipe, and chips of rock from the drill bit are brought up outside the pipe. In the bad old days, the chips were examined at the drill site, and bags of chips were stored. (This was how I first got interested in geology: when I went on day trips with my father, he gave me his petroleum engineer's explanation of the drill rig, the mud pumps, the drill pipe, and all the hardware. When I asked about the guys with the microscopes, he said, "Oh, those are geologists. Nobody understands them.") Drill chips are not much used today, although I have been tempted to build an automated X-ray diffraction machine for analyzing chips as they come out of the well. There is an active business called "mud logging," using instruments to detect traces of oil and gas brought up by the mud.

Drill chips are typically the size of a pea. Ordinary fossils are broken into unrecognizable chips. However, microfossils the size of a pinhead and smaller are intact within the drill chips. One particular fossil group, the Foraminifera, is especially useful. Some of the sandstones deep in the U.S. Gulf Coast have no surface equivalent; Latin names of their distinctive Foraminifera ("forams") are used for the sands: *Discorbis, Heterostegina, Marginulina.*

The small size of a foram collection once led to an interesting moral dilemma. When the government of Mexico ousted the international oil companies in 1938, one oil company paleontologist looked at his systematic foram collection, the key to the stratigraphy of the Mexican Gulf Coast. The entire collection was on the surface of a few microscope slides. He realized that he could destroy the entire collection in a few seconds by running his fingernail across the surface of the slides. As long as he left behind the slides and the crushed forams, he would have taken nothing physical away from Mexico. Was the organization and labeling of the collection his intellectual property? As a more practical matter: If caught, would he spend the rest of his life in a Mexican jail? As a historical fact, he did destroy the forams, he did escape safely, but the moral debate continued long after. It was a preview of the recent debate about downloading music from the Internet.

Even a capable geologist took many months to develop the skills for logging drill chips and years to learn forams. In one region in Kansas, some of the useful marker horizons were silty beds in otherwise fine-grained mudstones. Well-site geologists learned that chips of the silty beds felt gritty between their teeth. The roughnecks thought the geologists were identifying the rocks by tasting them. Even skilled practitioners of the chip-logging art produce slightly different inter-

Foraminifera are fossils, typically less than a millimeter in diameter. Because of their small size, they can be recovered from rock chips produced by the drill bit. © Ron Boardman; Frank Lane Picture Agency/CORBIS.

pretations from the same well. Something impersonal and objective was needed.

At the beginning of the twentieth century, two brothers, Pierre and Marcel Schlumberger, were searching for buried metal deposits using surface electrical measurements. They then realized that they could lower their electrical apparatus down a well on the end of a cable.[3] Their company, Schlumberger, came close to joining Thermos, Q-Tip, Band-Aid, and Xerox as a popular generic name for their product. Schlumberger today is not a monopoly; there are several competitors delivering almost comparable performance at lower prices. (As a kid, I thought the name should rhyme with "hamburger"; it's Schlum-ber-JHAY.) Schlumberger advertised itself as "the eyes of the oil industry." Initially, the Schlumberger company offered two types of downhole well logs: electrical voltages and measurements of the resistance to the flow of electrical current through the rock layers. In the new, scientifically correct world of SI units we are supposed to discuss "conductance," which is the reciprocal of resistance.

Over the years, Schlumberger introduced an impressive variety of downhole logging instruments. About 20 different logs are available today. Any objective measurement that you (or Schlumberger engineers) could think of can be measured at the bottom of a cable. In 1945, Schlumberger moved its headquarters from Paris to Houston. In 1960, before wine prices went sky-high, I could buy a Montrachet for $4 at a small wine shop in downtown Houston and get the additional thrill of being told, "Mr. Schlumberger loves that one."

Logs from objective, repeatable, downhole instruments found many uses. Subsurface layers were traced by matching up logs from adjacent wells, as if they were fingerprints. Sedimentary layers only a foot or two thick can be followed for hundreds of miles, particularly in sediments deposited on stable continental platforms. The opposite occurs in sands deposited by turbidity currents; wells 600 feet apart show little in common. Subtracting the depth of a feature on a downhole well log from the surface elevation at the top of the well gives a height above sea level. A structural map of folds and faults that affect a layer can be assembled from logs on a group of wells.

Another example: even on the simplest electrical logs, the difference between river-deposited point bar sands and an offshore bar is instantly identifiable. An offshore bar, with the coarsest sand on the top, looks like a cross section through a funnel. The pebbles and coarse sand at the bottom of a point bar sand make a shape like a bell.[4]

Each type of log measures something different. The first, *big* question asked is, "Where is the oil?" At the very beginning, the Schlumberger brothers realized that water, especially salt water, conducts electricity, and oil and gas do not. A high electrical resistance indicated oil or gas. That insight carried the Schlumberger company through its first 30 years. If the electrical resistance was low-low or high-high, there was no problem. Oil reservoirs are finicky; rocks containing more than 50 percent oil typically flow mostly oil; less than

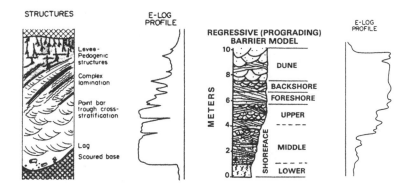

Point bar sands have their coarsest grains on the bottom; offshore bars are coarsest at the top. As a result, the wireline log response has the shape of a bell (point bar on the left) or a funnel (offshore bar on the right).

50 percent oil produces mostly water. In addition to the oil percentage, there were plenty of other attributes that affected the electrical resistance.

After 1920, most wells were drilled with the hole full of heavy mud to prevent the early-day blowouts and gushers. However, the heavy mud makes it possible to drill right through a productive oil reservoir and never know about it. An exact measure of the percentage of oil, or of natural gas, in a rock was desperately needed. The answer came, not from Schlumberger or its competitors, but from Gus E. Archie at Shell. Archie first published his method in 1941, but exploration for new oil was sidelined during World War II in favor of all-out production from existing fields.

In 1947, a Shell exploratory well at Elk City, Oklahoma, was about to be plugged and abandoned as a dry hole. Gus Archie argued that his calculations showed a zone that would produce oil. The company history records that "after some discussion" a test was made of the suspect zone.[5] It did produce oil; that well and the surrounding 136 wells in the Elk City oil field have produced 600 million barrels of oil and are still producing today. The 1947 discovery initiated an orgy. For the next few years, well logs from older "dry" holes were worked over with Archie's methods. Some of them did, in fact, indicate over-

Gus Archie (1907–78) developed a quantitative method for recognizing oil- and gas-saturated rocks from wireline well logs. Before 1947, some wells drilled right through oil-saturated layers without recognizing them. Archie's method is still widely used today. None of us at the time knew that his first name originally was "Gustavus." Photo courtesy of Shell Oil Company.

looked oil-saturated horizons. It was easy hunting; sometimes you could even drill out the cement plug and use the original borehole.

Archie actually discovered two relationships, sometimes called Archie's first law and Archie's second law.[6] His first law gives a relationship between the electrical resistance of a water sample and the resistance of a rock with its pore space saturated with the same water. His second law shows how the electrical resistance is changed as oil substitutes for the water. The two laws are combined into a simple equation known, obviously, as the Archie equation.[7] Gus Archie was my manager when I worked at the Shell research lab. It's great for morale to have a living legend on the team. Imagine being a rookie in the Yankee dugout and watching DiMaggio step up to the plate. There are lots of good scientists in the world and very few good managers of science. Archie was both.

Oil well logging tools improved after 1947, but the Archie equation is still in use today. When I was consulting on gas wells in New

York and Pennsylvania around 1980, I cranked out answers from the Archie equation for each one-foot interval and added up the answers to get the estimated gas reserves for the well. Long before 1980, photographic recording of well logs as wiggly lines had been replaced by computer processing. I asked whether the logging company could get its computer to run off the Archie equation for each foot and keep cumulating the result like an automobile odometer. The next time I was in upstate New York, the company had programmed an Archie odometer; all I had to do was subtract the odometer "mileage" from the top of an interval from the bottom "mileage." I could still use my judgment to pick the top and bottom of a productive interval, but I could get the reserves from a single subtraction.

During the oil- and gas-drilling booms of the 1970s and 1980s, Schlumberger generated an enormous cash flow. In the same way that the major oil companies found that diversifying out of oil was ineffective, Schlumberger also realized that "There's no business like *oil* business." Schlumberger diversified by acquiring other types of oil-service companies. In addition to its original wireline logging business, Schlumberger now offers "womb-to-tomb" services: exploration consulting, seismic crews, drilling rigs, well completions, and production programs.[8] If you have the money, you can outsource the operation of an entire oil company to Schlumberger. The two things that Schlumberger must never do is own its own wells or spread gossip.

At first, wireline downhole logs were kept confidential. Companies began trading copies of logs with one another, and eventually all logs were routinely released as public information. Some states, like New York, maintain public log libraries with photocopy machines. Other log libraries and commercial log-copying services exist. An individual geologist can make photocopies from the library for 10 cents each, spread the copies out on the kitchen table, and start making meaningful subsurface maps. If a plausible scenario for an oil trap can be worked up from the map, the geologist can sometimes lease the oil-drilling rights for a few dollars per acre. However, drilling a test well costs real money. The traditional investment formula is "a third for a quarter." Three outside investors each put up a third of the drilling costs, but they each get a quarter of the profits. The other quarter goes to the geologist. In those few instances when it works out, it is a quick exit from a dusty kitchen table to a mansion with a live-in chef.

In between the four remaining major oil companies and the individuals are a range of companies known as "independents." Actually, the larger independents are good-size corporations. The independents are the core of the American oil industry. Independents take most of the risks and find most of the new oil in the United States.

Subsurface geology works only where there are enough existing wells. In many places, the producing and dry wells are so far apart that big question marks turn up on the subsurface maps. Geophysics, a branch of the Earth sciences, can help. The most straightforward geophysical method, and one of the earliest, is measuring gravity. Every beginning physics student learns that the acceleration of gravity at the Earth's surface is 9.8 meters/second squared. As we say in the oil fields, "That's close enough for government work." It isn't exactly 9.8 m/s^2 everywhere. Altitude, the equatorial bulge, and subsurface rocks all make a difference. Salt domes rise up both because the salt can flow at depth and because the salt has a lower density than the surrounding rocks. A lower density means a slightly lower acceleration of gravity.[9] In the middle of the Avery Island pepper patch, the lighter salt beneath lowers gravity by 0.001 m/s^2, from 9.8 to 9.7999. Most first-year physics students wouldn't notice the difference, but a mass, delicately balanced against a calibrated spring, can find the salt dome. Unfortunately, salt domes are the only type of oil trap with a consistent gravity signature.

Gravity measurements have the disadvantage that the signal gets smaller as you go farther from the object; gravity fields diminish as the square of the distance. Twice as far away, a gravity signal is one-quarter as large. Ten times as far it's 1/100 the original size. Magnetic fields are even worse: they fall off as the cube of the distance. The only measurement that does not degrade with distance is the travel time of a wave. Radar and sonar measure the travel times of radio waves and sound waves. Radio waves don't penetrate much in rock, but sound waves travel through the entire Earth. The study of sound waves inside the Earth, called "seismic" waves, had a dual beginning. Detectors for seismic waves were just being installed around the world when the 1906 earthquake destroyed San Francisco. Studies of earthquake waves continue today. On a smaller scale, seismic waves were used by the Germans in World War I to locate Allied artillery batteries.[10] The locations were not exact; the Germans learned that sound waves traveled faster or slower on different paths through the ground. After the war a commercial venture, Seismos Gesellschaft, was founded. (The successor company is today a division of Schlumberger.) In 1924, Seismos crews began finding Gulf Coast salt domes by using explosives to send seismic waves horizontally through the rock.

Scott Petty, a founder of Petty Geophysical Company, gave something of the flavor of early seismic work:

> It would not be our first secret job as we had done it often before and we were well prepared and equipped. We had black gloves, long black rain coats, and black rain hats that buttoned around our faces and necks leaving only our eyes showing. Our Model T, our instruments and our tools were all black and our flashlights had black hoods over the lenses. Give us a good dark night on a country road and a drizzling rain and we were at home.[11]

During the 1920s a different seismic geometry was developed in Oklahoma. Sound waves, again from an explosive charge, traveled downward, were reflected from subsurface rock layers, and were detected as they returned to the surface. Major discoveries using this "reflection seismic" method began in 1930, and reflection seismology is the major workhorse today.

American physicists and engineers rapidly improved the bulky and balky early instruments. A portable seismic-wave detector was developed that is essentially an ordinary loudspeaker run backward. A loudspeaker runs electrical current through wires in a magnetic field to push the speaker cone. The small seismometer used the motion of the ground to push wires through a magnetic field to generate an electrical current. As a by-product of improving electronics, one seismic company, Texas Instruments, shared the invention of the integrated

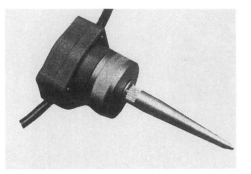

Detectors for reflection seismic work on land are properly called "geophones," but everyone calls them "jugs." "Jughustlers" are laborers who position and recover the detectors. One of my fellow jughustlers insisted that he was a "geophone retrieving engineer."

circuit with the founders of Intel.[12] Texas Instruments is today a major chip designer and manufacturer.

Until 1955, explosives were the sound source for virtually all seismic work. During the summer of 1952, I was working on a seismic crew, and for a while I was the assistant to the "shooter," the person who managed the explosives. One morning we loaded the truck with three tons of high-velocity dynamite, and as we bounced over the sagebrush, I asked what would happen if the dynamite exploded. "They find the engine block; some of the hubcaps."

During the 1950s, alternatives to explosives multiplied like rabbits. The winner that emerged was Vibroseis, patented by Conoco.[13] A Vibroseis unit looks like a compact, heavy truck. The truck lowers steel pads to lift the truck off the ground, and a hydraulic system shakes the entire mass of the truck to send seismic waves into the ground. A clever trick: the physicist Robert Dicke invented pulse compression for radar during World War II. The radar transmitter could not send out a huge pulse of energy all at once. Dicke had the radar send out a longer signal that started at a low frequency and swept gradually up to a high frequency. Then Dicke compared the returned signal with a copy of the transmitted pulse and found the exact time when the

A Vibroseis truck, with massive hydraulic loudspeakers, transmits sound waves into the ground.

two matched. In one sense, Dicke was not the inventor: bats and porpoises use similar "chirps" for sound ranging. So does Vibroseis.

A huge change in seismic methods occurred around 1955. Instead of recording on photographic film, seismic crews began to record on analog magnetic tape and then on digital magnetic tape. Computers could digest the data. Through the 1960s and 1970s, each time IBM or Cray brought out a bigger, faster mainframe computer, the major oil companies and the seismic service companies would each buy a half dozen. The physics behind the seismic method was fully worked out by 1919.[14] However, the calculation was what cryptography spooks call a "one-way function." If you tell me the distribution of rock properties in the ground, I can readily compute what the seismic waves will do. If you tell me what the seismic waves do, I have a computationally impossible task to reconstruct the rock properties. Running half a dozen IBM mainframes for the age of the universe won't

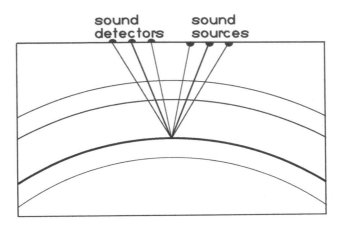

"Common depth point stacking" combines the seismic records from various sound source and sound detector locations to emphasize the response from each position in the ground. It is a method that works well in practice but is not easy to justify from pure theory.

come close. Seismic computations consist of tricks for getting usable approximations to the real answer.

Petty Geophysical gave up working on dark rainy nights and patented the most successful of the tricks. A seismic crew moves across the ground using many seismometers and many sound source locations. The trick involves summing, called "stacking," the signals at points halfway beneath each source and seismometer location. The method is called CDP, for "common depth point."[15] It works better than anyone expected; reflections appear like magic out of a mass of noise.

During the days of photographic recording, electronic amplifiers were adjusted to get some kind, any kind, of visible wiggle on the record. With digital recording, amplifiers could stay at a constant setting; the record would preserve the actual size of the signal. Seismic reflections from a sedimentary rock layer usually don't change much as you follow it along beneath the sagebrush. If the pore space changes from water-filled to oil-filled, the reflection changes a little. But putting natural gas in the pore space makes a huge difference: the gas is far easier to compress than oil or water. The abrupt increase of seismic reflection strength from the gas-saturated rock is called a "bright spot." A second clue: the bottom contact of the gas with the oil or water beneath is going to be exactly horizontal. The continuing success in finding oil off the shore of Louisiana and Texas is in large part due to detecting bright spots.[16]

The first 40 years of seismic reflection work were largely devoted to mapping subsurface geologic structures. In 1965, Shell Oil's success rate for hitting patch reefs in the Michigan Basin suddenly jumped from 2 hits out of 9 wells to 59 successes in 78 wells. Later, a published paper told how it was done. Gus Archie had assembled a "dream team": a geologist, a well-log analyst, and a seismologist. They found that one little wiggle on the seismic record changed over a buried patch reef.[17] They could locate a small patch reef, 30 feet thick and 3,000 feet down, by trusting that single wiggle. A new technique, "seismic stratigraphy," emerged. Reefs, shoestring sands, and porosity changes could be mapped using seismic reflections.

Usually, everything you do at sea costs more than on land. Reflection seismic exploration is a major exception. On land, picking up and moving the miniature seismometers is a major labor cost. Permission has to be obtained, and sometimes permit fees paid, to cross the land. At sea, a single ship can carry sound sources and can tow long strings of hydrophones to detect the reflections. Sound sources at sea are sudden releases of highly compressed air. Computers on board the ship do the initial processing.[18] Land reflection seismic work is 10 times more expensive than marine work.

In the past five years, the big push has been the introduction of three-dimensional seismic methods. Previously, everything was in two dimensions: the sound source and seismometers moved along a straight line. The CDP record was presumed to be a cross section beneath the surface line. However, reflections from features off the line ("side-swipes") were incorrectly included. In three-dimensional seismic surveys, the seismometers are spread out over an area, and the sound source successively visits locations within the area. Seismic crews expanded from 30 seismometers to 1,000 seismometers. Again, at sea it was much less expensive. The CDP stack became even more

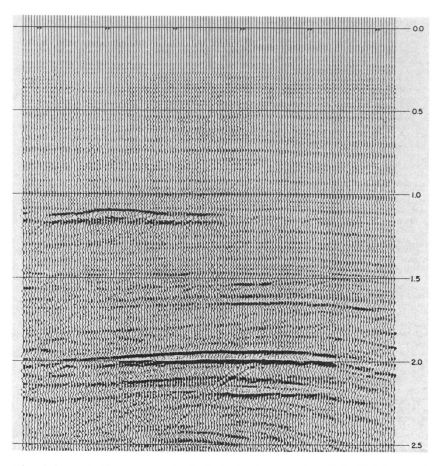

A local change in the amount of reflected seismic energy is called a "bright spot." Although oil and water have roughly similar seismic properties, natural gas is very different from oil or water because the gas is easily compressed. Not only is the natural gas valuable; the gas may be a gas cap on top of an oil accumulation.

effective; seismic paths were summed to a common depth point from all over the grid. There is plenty more to do. Some crews are using shear waves as well as compressional waves. Larger arrays of seismometers are being introduced. Clusters of hundreds of ordinary desktop computers are the new supercomputers.[19]

The enormous array of talent and money in the oil companies and the seismic service companies left most university scientists feel-

Ships used for reflection seismic work tow seismic detectors on long parallel cables. The sound is typically generated by a sudden release of compressed air in the water.

ing like outsiders, looking in with their noses pressed against the window. After James Cooley and John Tukey published a landmark mathematical paper announcing their discovery of the Fast Fourier Transform,[20] word leaked out that a seismic company in Calgary had been using the procedure for years. In 1970, however, Jon Claerbout at Stanford developed a novel way of processing seismic data, and companies provided financial support for his work.[21] In 1980, I was dealing with some highly unsatisfactory seismic data from a major geophysical company. Bob Phinney (Princeton) and John Costain (Virginia Tech)

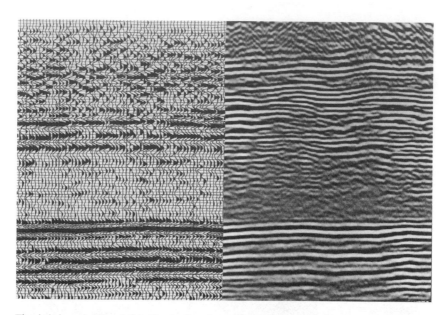

The left-hand side of this illustration is a reflection profile obtained by a leading seismic contractor. On the right is a profile at the same location measured by Virginia Tech, showing resolution of much finer detail. Before 1970, everyone presumed that the millions of dollars spent on research by major oil companies and large seismic contractors made it impossible for academic geophysicists to compete. It turns out that the universities do have something to contribute.

put together a high-resolution seismic survey that tripled the resolution of commercial survey.

The biggest academic help for seismic surveys came, not from a geophysicist, but from a structural geologist. It began with a practical question. The oil fields under the Alberta prairies continue west under the mountains. As oil company geologists drew cross sections through complex structures in the mountains, they realized that they had a problem. If they unfolded their cross sections, the rock layers could not be reassembled into flat parallel layers. In their cross sections, some layers were too short, others too long. Published cross sections in textbooks flunked the same test: all geologic cross sections of any complexity were demonstrably wrong. The geologists began working on geologic cross sections variously described as "balanced" (all beds the same original length) or "retrodeformable" (could be unfolded to a flat "layer cake" geometry).[22]

A Princeton structural geologist, John Suppe, picked up on the problem. He started by finding algebraic expressions for the dip angles in folds generated by fault motions. Gradually, he came to realize that almost all the folds we see in mountain ranges and oil fields are the indirect result of faulting.[23] Guided by those insights, Suppe assembled methods for drawing balanced or retrodeformable cross sections. Before long, he was a Big International Hero in the oil business.

John began traveling around to oil companies teaching one-week courses on his methods. I flunked Suppe's short course twice. Each time, for the first two days I was doing fine. I could find easier derivations for his equations and simpler ways to do the graphics. On the third day, John would say, "Now we guess at the answer and then we will show that it is correct." Hold it! How do you guess? Apparently, Suppe and his students draw on a mental collection of things that worked in the past and try the ones that might fit. It's a skill I simply don't have. That's why I teach sedimentation and John teaches structure.

Suppe's insights were particularly helpful in interpreting seismic cross sections. The user could infer things that were only hinted at by the seismic data. For instance, some styles of fault-induced folding tilt sediment layers until they are almost vertical. Vertical layers don't reflect seismic waves. Where the seismic images go blank, the informed interpreter thinks of vertical layers and tries to devise a Suppe interpretation that would include vertical layers.

Today, I get a weird feeling whenever a physician uses a sonogram to image my gizzard. Analog data, two-dimensional image, no CDP stack, no pulse compression: it's what we used in Wyoming in 1952. I'm just glad the doctor isn't using dynamite. It isn't trivially easy to translate modern seismic technology into medical practice. Deep seismic reflections in the oil patch take three and four seconds to return. In my body, everything is over in one millisecond. It will require fast computers and some tough engineers.

CHAPTER 5

Drilling Methods

The most powerful stimulant for finding more oil would be a reduction in drilling costs. Reservoirs that are now marginal could become economic. All those dry exploration wells would be a smaller drain on earnings.

If you started out today to design an automobile absolutely from scratch, you probably would not invent the gasoline-fueled internal combustion engine. The survival of the ordinary gasoline engine is not the result of some sort of conspiracy or conservatism. A hundred years of effort by backyard tinkerers, race car owners, and automotive engineers have produced a reasonably reliable and reasonably priced product. The rotary drill rig enjoys a similar place in the oil industry: a product of 100 years of gradual improvement.

The "spring pole," a simple human-powered drilling rig, can drill holes as deep as 1,000 feet. (The recommended "human power" is a couple of 220-pound Appalachian mountain boys.)[1] Drilling equipment in the nineteenth century was essentially a mechanization of the spring pole. A heavy steel chisel was bounced up and down on the bottom of the hole. Some water was added, and occasionally the drill was taken out and a cylinder called a "bailer" was lowered to remove the drill chips. Everything descended into the hole on cables; the rigs were called "cable tools." Because the hole was almost empty, discovery of oil or gas was obvious because oil or gas came out the top of the hole.

Smoking on a cable tool rig was absolutely forbidden; you can tell where a cable tool rig worked from the Copenhagen snuff cans in the surrounding bushes.

Encountering water raised a problem for cable tools. The bailer could not remove the water fast enough. The usual solution was to line the hole with steel pipe to shut off the water and then drill ahead with a smaller chisel that would fit inside the pipe. Each major water source required another string of pipe that would fit inside the earlier pipe. After several water layers, cable tools "ran out of hole"; the innermost hole became too narrow to continue drilling.

The U.S. Gulf Coast had oil, it had natural gas, and it had plenty of water-bearing layers on the way down. The "rotary rig" was the answer. Beaumont, Texas, has a memorial park celebrating the 1901 introduction of the rotary rig and the discovery of the Spindletop oil field.[2] The cable of the cable tool rig was replaced by steel pipe, with a toothed bit on the bottom. "Rotary" refers to rotating the pipe to make the bit cut into the rock. Mud was circulated down the drill pipe and back up between the pipe and the surrounding rock. The mud had several purposes: cooling the bit, bringing rock chips from the bit up to the surface, and holding back any water, gas, or oil in the surrounding rock. "Mud engineering" is an honorable profession, devoted to managing the viscosity, density, and filtering properties of drilling mud.

The major assumption behind the rotary drilling rig is having the pressure in the mud slightly higher than the pressure on the water, oil, or gas in the surrounding rocks. If the mud pressure is too low, then fluids from the surrounding rocks can either dilute the mud or, in the extreme case, blow all the mud out of the borehole. Mud pressures too high would cause mud to be lost into the adjacent rock. In 90 percent of all areas, the fluids in the rock have a predictable pressure; the subsurface fluid pressure equals the pressure due to a column of water extending up to the surface. Keeping the mud density in the borehole slightly larger than the density of water holds back the subsurface fluids without losing large amounts of mud into the surrounding rocks.

In 1938, engineers with Humble Oil (now ExxonMobil) reported
that wells deeper than 8,000 feet along the Texas and Louisiana coast
encountered fluid pressures 60 percent larger than expected.[3] "Expected" was the pressure due to a water column extending to the surface. The basic assumption underlying rotary drilling was challenged.
If the mud was made dense enough to control the fluids in the hole
below 8,000 feet, then the mud would be too dense for the layers shallower than 8,000 feet. Over the next 20 years, a technique evolved for
handling the problem: drill with normal-density mud to the top of the
high-pressure section, install steel casing to protect the low-pressure
part, increase the mud density, and drill on into the high-pressure section. The key step was recognizing the top of the high-pressure section
from wireline logs.[4] Just drilling ahead and waiting for the high pressures to announce themselves by blowing all the mud out of the hole
was the less attractive alternative.

These high pressures turned out to be significant both to understanding geology and to evaluating potential oil and gas resources.
Among geologists, Hubbert is more famous for pointing out the significance of abnormally high pressures than he is for predicting the
peak in U.S. oil production. In 1959, Hubbert published two papers
coauthored with Bill Rubey.[5] Inadvertently, Hubbert and Rubey
launched a thousand spoonerisms. After one speaker at a meeting referred to "Rubert and Hubby," Hubbert prefaced his question to the
speaker by saying, "I'm Hubby."

The puzzle that attracted Hubey and Rubbert's attention occurred in the Alps and in other great mountain ranges. Slabs of rock,
with horizontal dimensions something like 50 by 50 miles and a mile
thick, apparently had traveled over the rocks beneath. The puzzle was
mechanical. Suppose you give me an enormous parking lot, and on
top of the parking lot you place a slab of good hard rock 50 miles by
50 miles in area and a mile thick. You also loan me some giant bulldozers to push the rock slab across the parking lot. I couldn't. Although the bulldozers could push hard enough, before they could
overcome the friction between the rock slab and the underlying parking lot, they would crush the back end of the rock slab. The rock isn't
strong enough to be pushed around.

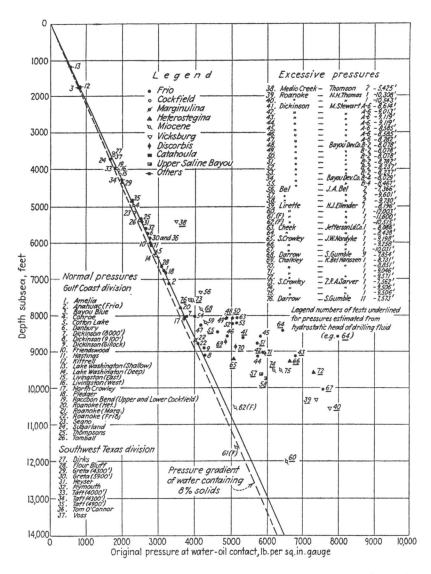

An abnormal increase in pressure below 8,000 feet in Gulf Coast wells was first reported in 1938. © Society of Petroleum Engineers.

Hubbert and Rubey pointed out that high fluid pressures, as found beneath the Gulf Coast, were the key to the puzzle. The friction between the dry rock slab and the dry parking lot is generated by the weight of the rock slab. Hubbert and Rubey explained that if water, under pressure, exists between the rock slab and the parking lot, then

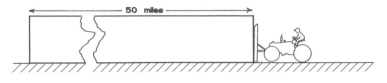

A hypothetical, but impossible, mission: slide a 50-mile-wide block across a parking lot. The block would break up in front of the bulldozer blade instead of sliding.

the friction is less. It is not that the water is lubricating the slab; it is not greasing the parking lot. If the water pressure holds up part of the weight of the overlying slab, then friction is due only to the portion of the slab weight that is *not* held up by the water pressure. In the extreme case, if the water pressure equals the entire weight of the rock slab, then I wouldn't need the bulldozers. I could push the rock slab around with one hand. (My son reports seeing a crew rearrange a store interior by tipping the sales counter, slipping a plate beneath, turning on a pump that blew air out through perforations in the bottom of the plate, and then pushing the assembly around by hand.) Many accounts of great rock slides begin: "It rained hard the night before." Filling up cracks and pores in the rock with rainwater raises the pressure just enough to trigger a rock avalanche.

Hubbert and Rubey used the Greek letter σ (sigma) for the pressure due to the overlying rock *minus* the water pressure, which they called the "effective stress." Normally, the water pressure is the pressure due to a column of water all the way up to the surface, and the rock pressure is due to the rock column. Typically, the specific gravity of the rock is 2.2, and the specific gravity of water is 1. The water pressure is about 45 percent (1/2.2) of the total pressure; the other 55 percent is the effective stress held up by the solid rock. This situation is described as "normal pressure" because it describes about 90 percent of the places where wells are drilled. "Overpressure" refers to places, like the Gulf Coast subsurface, where water pressures are higher and the effective stress is less than 55 percent of the overlying rock column.

Overpressures, if high enough, constitute the heavy magic that allows enormous rock slabs to move in major mountain ranges. The Gulf Coast itself displays an interesting geologic expression of the overpressures. Earthquake hazard maps show the Gulf Coast with zero earthquake probability, the lowest in the country. Yet there are abundant Gulf Coast faults that offset young sedimentary layers, usually in the direction that moves units toward the gulf. One of the faults offset the pavement in the old Houston-Galveston road by six inches.[6] The faulting is happening today without generating historic earthquakes. The Gulf Coast is quietly "slip sliding away" into the Gulf of Mexico. The cause of overpressuring is another of those problems that generate far too many clever explanations.[7] I am convinced that reactions between water and water-bearing minerals are the major cause. Here's my pitch.

If both liquid water and a mineral that contains water are present, the water and the mineral have to be in equilibrium. Equilibrium requires that water lost from the mineral have the same energy (called the "chemical potential") as the liquid water. Temperature and pressure change the chemical potential for both the water and the mineral.[8] This is best shown on a graph. (The graph is *not* to be confused with an Arrhenius diagram, which has a similar appearance.) The vertical axis is the effective stress, σ, as used by Hubbert and Rubey. The horizontal axis is the temperature. Because the line describing a mineral dehydration slopes from upper left to lower right, increasing temperature causes the mineral to say "I can't take the effective stress anymore" and hand off part of the overburden load to the water.

It helps to discuss a specific example. The mineral gypsum contains two water molecules and one molecule of calcium sulfate. Anhydrite contains just calcium sulfate. Gypsum is a mineral commonly deposited at the surface by evaporation of seawater. At a burial depth of about 1,000 feet, gypsum starts changing to anhydrite by giving up the water molecules.[9] If there is a connection to the surface through porous sandstones, or through leaky fractures, the water can leave, and the gypsum layer becomes an anhydrite layer. On the other hand, if there is an impermeable barrier around the gypsum layer, then the

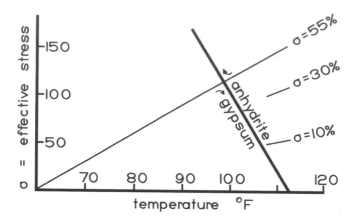

The heavy line shows the boundary between the stability field of gypsum, $CaSO_4 \cdot 2H_2O$, and the stability field of anhydrite, $CaSO_4$. The thin line labeled $\sigma = 55\%$ indicates the conditions expected if the water pressure is determined by a continuous column of water extending to the surface.

mineral-water assemblage has to follow the gypsum-anhydrite boundary. As the temperature increases with further burial, the effective stress moves toward zero. As Hubbert and Rubey point out, it is the effective stress that determines the resistance of the rock to sliding along faults. If water cannot leave and the temperature continues to increase, the rock exhibits extremely weak behavior. North of the Alps, the Jura Mountains look as if they slid on a bed of gypsum that was changing to anhydrite.[10] (The same mountains gave their name to the Jurassic period in the geologic time scale.)

What happens if erosion brings a deeply buried anhydrite bed back to the surface? Again, if water has access to the rock, it converts back to gypsum. However, deeply buried beds frequently have lost their pore space; the permeability may be very low: suitcase rock. In the absence of an adequate supply of liquid water, the rock will again follow the gypsum-anhydrite boundary, but this time lowering the temperature will bring on increasing effective stress and zero water pressure. To put it bluntly: the rock sucks.

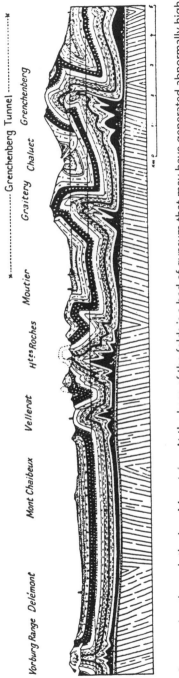

Cross section through the Jura Mountains. At the base of the folds is a bed of gypsum that may have generated abnormally high water pressure.

The mineral involved in Gulf Coast overpressuring has to be something other than gypsum. Most of the sediments were deposited as mud, and the dominant mineral in the Gulf Coast mud is a clay mineral called montmorillonite. (The more proper group name is "smectite," but smectite always looked to me like a word that belonged in a sex manual.) Most clay minerals are silicate sheets with layers of water molecules in between the silicate sheets.[11] At room temperature and 60 percent relative humidity, montmorillonite contains two layers of water molecules. Raise the temperature, or lower the relative humidity, and montmorillonite loses water to convert to a one-water-layer structure. At temperatures above 300°F, montmorillonite loses the last of its water layers. Losing water from montmorillonite is not as abrupt as the gypsum-anhydrite transition. Instead of the abrupt chemical reaction from two-water gypsum to no-water anhydrite, montmorillonite has gradual changes. On a graph of effective stress against temperature, the water percentage in montmorillonite can be shown as contour lines. However, the contour lines pack closer together at the boundary between the two-water-layer clay and the one-water-layer clay. Instead of a sharp boundary as in gypsum-anhydrite, there is a fuzzy boundary. If there is no permeable connection to the surface, when the mud reaches the temperature associated with changing from two water layers to one, the mud will become overpressured. The mud itself can act as a low permeability seal. I am quite convinced that loss of water from montmorillonite clay causes the Gulf Coast overpressuring.

In the eventual limit, the effective stress can go to zero, and the rock will exhibit no strength at all. An offshore Texas well apparently drilled into the zero-strength material. The drillers knew they were in severely overpressured rock and had their mud density up almost to the density of average Gulf Coast rock. Drilling became easier and easier, and they discovered that they could stop rotating the bit, keep pumping down heavy mud, and just lower the drill pipe in the hole. After a few hundred feet of penetration, they realized that this might not be A Smart Thing. If they pumped their way into a gas-filled sandstone, they might all die in the ensuing blowout. In their newfound

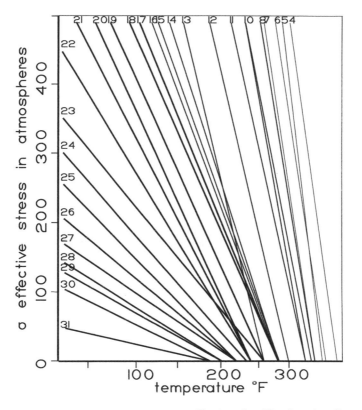

Weight percentage of water in montmorillonite clay. The heaviest lines correspond to clay with two water layers, medium lines to single water layers, and the thinnest lines to zero water layers.

wisdom, they decided to pull the drill pipe out and plug the hole with cement.

There are a number of known overpressured provinces around the world;[12] the Gulf Coast is not unique. Taiwan is particularly interesting. Whereas the Gulf Coast is sliding gently down toward the ocean, Taiwan is being pushed into a newly formed mountain range. There is natural gas (but little oil) on Taiwan. In central and western Taiwan the subsurface effective stress is supporting only about 25 percent of the overlying rock[13] (as compared with 55 percent in the normally pressured rocks). Southernmost Taiwan gives the impression of being fully overpressured, much like the offshore Texas well. Mud "volcanoes" rise up from depth to the surface; bubbles of natural gas come up with the mud. Some puzzling chaotic structures in ancient mountain ranges may be rocks stirred by ancient mud volcanoes.[14]

The temperature of clay dehydration, and the onset of overpressuring, is near the base of the oil window. In the Gulf Coast, and on Taiwan, most of the hydrocarbon within the overpressured sections is natural gas.[15] A limited amount of oil and larger amounts of gas condensate are present near the top of the overpressured rocks. "Dry" natural gas, almost pure methane, makes up most of the hydrocarbons in the overpressured sections. Over the next 10 years, our oil supply problem is not going to be solved by exploring the overpressured rocks. On the other hand, natural gas and gas condensate from overpressured rocks will be an important source of both fuel and raw materials for petrochemicals.

Drilling, either in normally pressured or in overpressured rocks, is accomplished by a drill bit at the bottom of the pipe. Out of the Spindletop experience came a line of drill bits with rotating toothed cones. The most successful bit manufacturer was the Hughes Tool Company, founded by Howard R. Hughes.[16] His son, also named Howard, was able to pursue his adventures with movies, electronics, exotic aircraft, and exotic actresses because he inherited Hughes Tool Company, a solid cash cow.

Rotary drillers learned that a rolling, toothed bit, when it was working at its best, returned chips of rock the size of a driller's dirty fingernail. Wells in the soft sediments of the Gulf Coast were drilled as fast as 1,000 feet per day. However, drilling through a previously depleted natural gas sand, 40 feet thick, on the Gulf Coast could take a week and wear out three bits. It was the difference between chopping through a log by turning out chips from a sharp ax and bashing through the same log with a sledgehammer. The gas sand was still as loose as the sand in a playground sandbox, but the high pressure in the mud and the low pressure in the depleted gas reservoir turned a chipping process into a grinding process.

During the twentieth century, rotary rigs underwent a kind of Darwinian evolution. Rigs with the most powerful mud pumps seemed

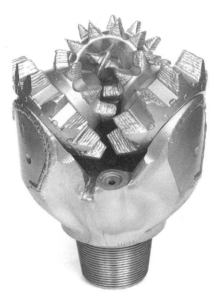

The tricone drill bit was developed by Howard Hughes's father, beginning around 1908. Engineering the bearings at the centers of the movable cones was at least as difficult as designing the cutting teeth. Photo courtesy of Hughes Christensen.

to drill faster. Larger and yet more powerful mud pumps were purchased. Rotating the drill pipe and drill bit required 30 horsepower, but the mud pumps consumed 2,000 horsepower. It paid off in faster drilling, but why? The answer is found in the Bernoulli equation,[17] developed by a member of a Swiss family of notable mathematicians. In a flowing fluid there are three contributions to the fluid energy: pressure, velocity, and height. The Bernoulli equation is a statement about conservation of energy: move the fluid faster, and the pressure has to decrease. The fast mud pumps lowered the pressure around the bit; chips came flying off the bit teeth.

The Bernoulli effect, lowered pressures caused by increased velocity, is not just an exotic effect in the bottoms of oil wells; it explains how airplanes fly. An airplane wing is relatively flat on the bottom, but the wing has a curved upper surface. Air flows straight under the wing with no change in speed, or pressure. Above the wing, the air has to speed up in order to detour over the curved upper surface. An increase of speed lowers the pressure over the wing enough to lift the airplane. I recommend changing the song title: "You are the wind *over my wings.*"

In earlier days, a few of the drilling rigs were owned and operated by the major oil companies. Today, all drilling is done by contractors. A well a few thousand feet deep costs about $25 per foot. (My own career as a drilling contractor never got off the ground. When I was little, wells were drilled for $3 per foot. I offered to hire out with my little shovel and pail and dig the first foot.)

Oil well disasters have the potential to be major environmental disasters. People are not as tolerant as they used to be about being sprayed with crude oil for days. An important key to safety is an unimportant-sounding item known as "surface casing." A well is initially drilled down into solid rock, and large-diameter surface casing is lowered into the hole and cemented in place. Typical surface-casing depths are a few hundred feet. Atop the surface casing are mounted giant hydraulically operated valves, known as "blowout preventers." If a well starts to flow uncontrolled amounts of oil or natural gas, crew members quite understandably run for their lives. Right at the bottom of the steps leading away from the rig are big red buttons; on the way by, someone gives a whap to a red button, which closes the blowout preventers. If that happens, the only grip the system has on the ground is the cement bond between the rock and the surface casing. If either the cement bond or the rock is weak, then oil and gas start coming up around the borehole and create a surface crater. By extension, to "crater" became a verb in 1950s oil field slang for any disaster.

I have never forgiven Union Oil of California for its handling of a 1969 well in the ocean off Santa Barbara.[18] Major amounts of oil leaked around the surface casing, and Union Oil explained that its surface casing met the federal guidelines. My own outlook says that the petroleum engineer has the responsibility of specifying and installing an adequate system and that the government is only a sideline participant.

For a law firm in Michigan, I looked at well records from a different kind of environmental disaster. A deep well had encountered high-pressure gas, including some toxic hydrogen sulfide. Drinking-water wells on nearby farms started bubbling gas. For most consulting projects, I usually volunteer to take an initial peek at the data without payment. If it looks as if I can do some good, then I begin working for money. After spending one day going over the well logs, I wrote a note to the lawyer saying that it was obvious that the surface casing wasn't deep enough to protect the farmers' water horizon. I never heard from the lawyer again. I bet he showed my note to the oil company, and it decided to settle.

Hydrogen sulfide must be removed from natural gas before it goes to market. The hydrogen sulfide can be converted to sulfur. In fact, sulfur as a by-product from purifying natural gas has driven most other sulfur sources out of the market. Although people joke about the rotten-egg smell, hydrogen sulfide is as toxic as hydrogen cyanide. Most natural gas does not contain hydrogen sulfide, but a few areas produce gas with 5 to 10 percent of the rotten stuff. Working around those wells requires one person to put on a gas mask and do the work and a second person to stand by with a gas mask to rescue the first guy if he passes out. The nastiest possible environmental disaster is to have a gas well, with hydrogen sulfide, blowing wild and out of control. It has happened. The first required step is to light the gas, to start the well burning. (Not with a match; use a tracer bullet from a rifle.) The hydrogen sulfide burns in air to form sulfur dioxide. Sulfur dioxide isn't all that nice, but it isn't instant death.

Oil and gas fires are scary, big-time scary. I was enormously surprised that the retreating Iraqi troops set fire to the oil wells in Kuwait. An ancient tradition among desert tribes prohibits the poisoning of water wells, even against one's vilest enemies. The Kuwait fires were predicted to take years to bring under control, but American and Canadian crews did a remarkable job of getting the fires out and the wells under control in a few months. Big alpha males are macho, bullfighters have machismo, machissimo is reserved for oil well fire fighters.

After a well is drilled to its planned depth, a logging company (Schlumberger or one of its competitors) makes several records by lowering various instruments into the hole. Examination of the logs, usually using the Archie equation, shows whether there are zones likely to produce oil or gas. At this time, there is another tough decision. Drilling the well is typically 50 percent of the expense; the other 50 percent is the "completion" cost of preparing the well for production. Typically, six hours is allowed for the investors to be consulted about participating in the cost of completion. The six hours always seem to begin right after midnight. A great deal of mutual trust is involved in discussing the logging results and making oral agreements to participate financially in the completion. The oil business has its own internal moral code. Sinners are ostracized; I refuse even to talk on the phone with anyone who I think broke the unwritten rules.

There is an intermediate, but costly, step between the downhole well logging and completion of a producing well. The drilling rig and its drill pipe can be used to produce the well for a few hours. The tactic is known as a "drill stem test," drill stem being a synonym for the drill pipe. A drill stem test can produce oil, gas, water, or nothing. ("Nothing" means you were trying to produce out of suitcase rock.) Here is an example of how difficult it is to keep a secret in the oil business: a major company drilled an exploratory well in partnership with a small Canadian company. The Canadian partner, however, was big enough to have its stock traded on a stock exchange. People waited anxiously as the well reached its expected depth and was logged, but there was no public release of information. However, the safety engineer for the major company backed his company pickup truck down his driveway at 3 A.M. Observers knew that company regulations required the safety engineer to be present to oversee the hazards of a drill stem test. That meant that the downhole logging results had to be good enough to justify a drill stem test. Phones rang all over North America. Success would not make a dent in the major partner's earnings; but when stockbrokers arrived for work later that morning, they found a pile of "buy" orders for the Canadian partner's stock.

The first stage of a well completion is running steel casing, typi-cally seven inches in diameter, to the bottom of the well and anchor-ing it in place with cement. (In the oil patch, it's SEE-ment. I know, I know, the dictionary says se-MENT.) On the main street of Duncan, Oklahoma, is a statue of Erle P. Halliburton, a pioneer in methods for pumping cement behind the casing.[19] His company still has a major presence in Duncan, although Halliburton offers services worldwide and now has its headquarters in Dallas.

After the cement has sealed the casing to the surrounding rock, shaped explosive charges blast holes through the casing. These holes are carefully located in the zones where the downhole logs indicated oil. The well now has 6 to 12 holes, each one inch in diameter, con-nected into the oil-bearing rock. Around 1960, a technique was intro-duced that greatly improved the connections between those holes and the surrounding rock. The truck-mounted pumps used by Halliburton and its competitors could generate fluid pressures high enough to crack open the rock at depth.[20] The process was originally named "hydrofracturing," quickly shortened to "frac." Gradually the frac process was improved with gels, with foam, and especially with the introduction of sand-size grains to prop open the newly formed frac-tures. Numerous formations that would not have produced economic amounts of oil or gas became viable because of the frac process.

Here is an exercise for the creative reader: long before the frac process was invented, wells were "shot" by exploding cans of pure liq-uid nitroglycerine in the well. Besides being incredibly dangerous to arrange, the nitroglycerine explosion happened in a thousandth of a second and simply crushed the rock and pushed it back from the well bore. At the other extreme, a conventional frac job takes minutes to hours and opens one long, single fracture in the easiest direction. The ideal procedure would open about six fractures radiating in different directions from the well bore. About 20 years ago, the bomb jockeys at Sandia Labs ran this problem through their nuclear weapons com-puter codes and concluded that you would get roughly six fractures if you could supply the energy over one second. One second, in between the nitroglycerine millisecond and the one-hour frac job. What takes one second? After some scratching of heads, one of the nuke nerds remembered that gunpowder for the largest naval guns comes in inch-size granules that take about one second to burn. They conducted a satisfactory test underground at the Nevada Test Site, but the process never became competitive in the oil fields.[21] I was told that it was not possible to stuff enough naval gunpowder to do the deed in the limited borehole volume. The challenge: devise a way to deliver a large volume of high-pressure fluid, spread over one second. Don't call me. Call Halliburton at 214-978-2600, Dowell at 212-350-9400, or BJ at 213-462-4239. When you test it, I want to be far, far away.

Once a well is completed and producing, the well operator would love to sit back, relax, and take checks to the bank. No such luck. The better wells initially will flow oil and gas spontaneously, but after a while the pressure at the bottom of the hole will no longer lift oil to the surface. Time to install a pump. Another useful lesson from beginning physics: even a perfect pump will only lift a water column 30 feet high. Pumps for deeper wells have to be located at the bottom of the hole. Those rocking devices that you see in oil fields are lifting steel rods (known, inelegantly, as "sucker rods") connected to a pump at the bottom of the well.

Natural gas separating from the oil acts like a refrigerator, and cooling can cause paraffin or asphalt deposits to separate from the oil and clog up the equipment. One of my more interesting consulting adventures was with a New Jersey company that developed materials for the printing industry. Greasy ink is removed from printing presses by washing with a gasoline-like solvent. The group in northern New Jersey developed a mixture that could be added to the ink at cleanup time; the ink could then be washed off with water. One serious problem: the cleaning agent made the press rollers so water-loving that it was impossible to get fresh ink back on the press. Someone suggested, apparently sarcastically, pouring the new goo down oil wells to clean out the paraffin and asphalt. It worked! The goo has been patented and is being sold to oil well service companies.[22]

The success of an oil field depends on the energy available to drive oil from the reservoir rock into the wells. There are three sources of energy,[23] but they differ enormously in effectiveness:

1 If the oil reservoir has a natural connection to water, as in an ordinary anticline, then water can displace oil out of the reservoir rock and into the wells. The efficiency of this "water drive" can drive as much as 60 percent of the original oil into the wells.

2 If there is initially a separate accumulation of natural gas above the oil, the operator can locate the casing perforations only in the oil-saturated zone. As the oil is produced, the gas cap expands to displace oil, but the process recovers only about 40 percent of the oil.

3 In most oil reservoirs, initially there is some natural gas dissolved in the oil. As the oil field is produced, bubbles of gas separate from the oil and push oil toward the wells. Typically, less than 20 percent of the oil is recovered.

These low recovery efficiencies inevitably raise the question: Can we recover the rest of the oil? The first answer was found by accident. Oil well operators during the nineteenth century noticed this: wells leaking water from a shallow horizon down into the oil reservoir often improved the production from nearby wells. Some operators began pouring water down some of their wells in order to get better oil production from adjoining wells. This was often regarded as cheating. The New York legislature passed a law making it illegal to pour water down an oil well. Of course, it turned out that the artificially introduced water had the same effect as a natural water drive. Water flooding became a common tactic by 1950. Even New York relented.

Although a gas cap expansion is not as efficient as a water drive, it can be an inexpensive conservation measure. All too often, natural gas produced with oil was burned in enormous flares. Injecting the produced gas back into an existing gas cap in the short run aids production; in the long run the gas can be marketed after the oil production ends. Engineers in one oil field joke that they have pumped gas back through the reservoir so many times that they have worn all the corners off the gas molecules.[24]

Water floods and gas injection were called "secondary recovery": stopgap measures to be employed after "primary recovery" became uneconomic. Gradually, "secondary" recovery was employed earlier and earlier. In the modern approach, techniques that once were "secondary" are started with the very initial production. As an example, early water injection can be used to prevent gas bubbles from forming in the oil.

Planning the entire life of the oil field soon after the field is discovered has substantial benefits. Petroleum engineers can locate water injection wells and gas injection wells as part of the plan for drilling the early production wells. The program can be optimized for total oil recovery instead of maximizing the early cash flow. Computer simulations allow "what if" questions to be asked while planning the field. However, simple computer simulations can be highly misleading. The computer problem is not highly difficult for a natural gas reservoir in a mythical "homogeneous isotropic" reservoir with no natural water drive. Real reservoir rocks have lots of internal structure. For instance, little streaks of mudstone a quarter inch thick have a huge effect on fluid flow. The rate of flow of gas, oil, and water is highly dependent on *how much* gas, oil, and water are present. At the very birth of the modern computer age, John von Neumann identified weather forecasting and oil reservoirs as important problems requiring huge amounts of computer power.[25]

Even at best, an effective water flood will recover 60 percent of the oil. Can we get at the rest? There are several techniques lumped together as "tertiary recovery" or "enhanced recovery," methods used after secondary recovery has been finished.[26] Here are some examples:

1 Steam floods, injecting steam to warm up thick viscous oil, to melt asphalt and paraffin deposits, and to move the oil along with the condensed water. Steam floods exist both as steady injections and as "huff and puff" operations that inject steam into a well and then pump oil right back out of the same well.

2 Detergent floods, injecting small amounts of detergent into flood water to mobilize oil as very small droplets. Detergent floods have been successful in some shallow oil reservoirs. Deeper reservoirs are often hot enough to break down the detergent chemicals.

3 Fire floods, injecting air and starting a fire inside the reservoir. The air is limited in amount so as not to burn the entire reservoir contents; the hot combustion gases push oil from injection wells to the recovery wells. One serious problem: burning creates some strong chemicals. Sulfur in the oil can wind up as sulfuric acid. A fire flood can be less than economic if the sulfuric acid eats the steel pipe out of your recovery wells.

4 Miscible floods, using liquefied gases such as butane and propane to solvent wash the oil out of the reservoir. This is the most efficient method of all, with recoveries approaching 100 percent. Unfortunately, it is applicable only in special situations. The butane and propane are expensive; owners of gas grills and recreational trailers are willing to pay good money for the same gas. Of course, the oil field operator counts on marketing the butane and propane after the flood is complete, so the reservoir better not leak either in or out. Miscible floods are risked only where the reservoir is demonstrably isolated from water in the surrounding rocks.

5 Carbon dioxide floods, using either carbon dioxide gas or carbon dioxide dissolved in water. Carbon dioxide is soluble in water; it is even more soluble in oil: it increases the volume of the oil and promotes oil migration into the recovery wells. At the recovery well, the carbon dioxide is separated from the oil and the carbon dioxide sent back around to pick up more oil. Several petroleum geologists have put their hats on backward and drilled successfully for carbon dioxide for flooding projects.

There is no single magic bullet. Each of the advanced recovery techniques is economic under the appropriate circumstances. One sorting technique developed by the U.S. Department of Energy listed the special circumstances appropriate for each method, but any reservoir that flunked all the other criteria was automatically listed as a candidate for a carbon dioxide flood. There is talk of modifying electric generating plants to keep their carbon dioxide out of the atmosphere. If it happens, the oil fields would be happy to use the carbon dioxide.

All natural gas, and most oil, are transferred to pipelines. (Tank trucks carry away small amounts of crude oil.) Originally, pipeline companies bought oil and gas at the wells and sold oil to refineries and gas to consumers. Pipelines, and especially gas pipelines, are increasingly being treated as common carriers. An owner of natural gas wells and a consumer of gas can agree on a price. The pipeline gets a fixed fee for hauling the gas. As long as the gas meets specifications, the consumer does not have to receive the actual gas molecules delivered by the well owner. A similar arrangement is becoming possible for electricity; a homeowner can pay extra for "green" electricity from non-polluting and renewable sources. The power company doesn't paint the electrons green to make sure they are delivered to the right house.

Progress continues on better drilling methods and better methods of producing oil and gas. A theme throughout this book: progress has been going on for a long time, and there is little expectation that something dramatic will come riding to the rescue as world oil production starts to decline. On drill bits, tungsten carbide buttons replaced steel teeth. A downhole hammer, a bigger version of the construction-site jackhammer, drills faster when the rock is hard and the well is less than a mile deep. Fracturing methods use ever more exotic fluids.

The most recent twist in drill bits is almost a throwback to the "fishtail" bits of 1910. The new version is called a "diamond compact bit." Instead of having teeth mounted on a rolling cutter, the diamond compact bit has fixed teeth, each tooth faced with 1/16 of an inch of manufactured diamonds.[27] Sometimes an entire well, 6,000 feet deep, can be drilled using a single diamond compact bit. The diamond bit costs $100,000, as opposed to $10,000 for a toothed-cone bit; the economic advantage comes from saving time and labor in pulling out the drill pipe to change bits.

Diamond compact bits have fixed teeth armored with a layer of fine-grained diamond that has been squeezed and compacted into a solid layer. Photo courtesy of Hughes Christensen.

Another innovation began 10 years ago as a unit used for servicing existing oil wells. Instead of screwing together lengths of steel tubing to reach to the bottom of a well, several thousand feet of continuous steel tubing was rolled up on a spool about 50 feet in diameter. These big contraptions are called "coil-tubing rigs." The tubing could be reeled rapidly into or out of the hole. Because the outside of the tubing was smooth, with no threaded joints, a rubber wiper allowed the tubing to go into the hole even with the well flowing oil or gas to the surface.

During the past two years, entire new wells have been drilled from the surface using coil-tubing rigs.[28] The bit is rotated by a downhole turbine, operated by the mud pressure. Because the tubing on the coil-tubing rig is rather narrow, two or three inches in diameter, the well has a small diameter. Early experiments, conducted in Oman, were used to reach bits and scraps of the reservoir that would not be drained by the larger wells. About every 30 years, "slim-hole drilling" gets reinvented. Usually, the industry goes back to seven- or nine-inch

This coil-tubing rig consists of a continuous length of steel tubing wrapped on the drum on the right plus guides to unreel the tubing into the well on the left. Photo courtesy of Dowell-Schlumberger.

holes because mud flows better through the larger system. The next evolutionary step might be to even larger coil tubing. The drum would look like a Ferris wheel, but big size has seldom been a barrier in the oil patch.

Of the new techniques, directional drilling and horizontal drilling come closest to having the U.S. Cavalry gallop into sight. Originally, a crooked hole was bad news: when the well was pumped, the sucker rods would wear against the pipe. A lot of wells, particularly those drilled in inclined sedimentary layers, were not straight. Ownership of the oil in the ground was supposed to extend straight down beneath the surface. A well that produced from beneath an adjacent property was "crooked" in two different senses of the word. (I saw one map hinting that a deep gas well in eastern Austria actually produces from beneath the Czech Republic.) As instruments were developed for surveying boreholes, serious lawsuits arose. In some oil fields, especially in California on steep-sided anticlines, all the wells had unintentional wanderings. The state of California instituted a moratorium banning directional surveys of older wells. All existing wells were grandfathered as legal.

Directional wells were useful even where steeply dipping layers were not present. A well could produce from beneath surface houses or a lake. Offshore, 20 wells could be drilled in different directions from a single fixed platform. Wells penetrate as much as three miles horizontally without ever going more than a mile beneath the surface. As the techniques got better and better, a very old question was raised once more. Producing wells, typically 600 feet apart, going straight down through a nearly horizontal oil reservoir amount to one square foot of hole for a million square feet of reservoir. Could you drill a horizontal hole through the reservoir? Yes. Horizontal drilling has become a standard feature for oil and gas production. In particular, some of the major fields in Saudi Arabia have been redrilled with horizontal holes to obtain better recovery.[29]

The gadgetry for horizontal drilling is amazing. Originally, steel wedges were placed to deflect the drill. Today the bottommost foot of the drill pipe, just above the bit, is kinked out of line by about 10 degrees. Instead of rotating the drill pipe to rotate the bit, circulating mud goes through a bottom-hole turbine to turn the bit. (The downhole turbine is a Russian invention.) If you want to drill straight, rotating the drill pipe slowly from the surface causes the 10-degree kink to average out. If you want to curve the hole, stop rotating the drill pipe, and the kinked bottom section will start drilling around a curve.

Horizontal drilling is made more efficient by placing downhole sensors above the bit.[30] Some of the measurements made by wireline logging after the hole was drilled can be made while the well is drilling. The driller can navigate the well through the best part of the reservoir by following the downhole sensors. In the new world of horizontal drilling, the driller sits in an upholstered barber chair facing a bank of computer screens with joystick controllers in both hands. It's like the world's biggest video game. Don't worry, before long the driller will be operating the rig over the Internet from a Houston high-rise, with clean fingernails. Kids: the best training for this job is playing computer games.

CHAPTER 6

Size and Discoverability of Oil Fields

Would we find more oil if we simply drilled holes at random? I had hoped that the answer would be a loud "No!" In 1975, however, Bill Menard of the Scripps Institution of Oceanography suggested that the answer was a quiet "Yes." Menard calculated that we would have found the East Texas oil field (7 billion barrels) much earlier if we had been throwing darts at a map.[1] In fact, the odds were better than a billion to one that East Texas would have been found earlier by drilling at random. My own reaction was that Dad Joiner *did* discover East Texas by random drilling. Joiner used a hunch from a local veterinarian, which is pretty close to random. However, Menard's discussion needed a closer look. He knew the oil business; he organized the first underwater-mapping team.

When Menard's paper was published, I was starting to work with statisticians at Princeton on oil field data. An old saying in statistics: "You don't have to eat the whole ox to know the meat is tough." Rather than trying to digest all the world's oil fields at once, we needed a manageable example. I suggested Kansas for two reasons:

1 I had heard that there were 30 "one-well oil fields" in Kansas. You know that you are scraping the bottom of the barrel when you find the one-well oil fields. An oil field with only one well is usually an economic disaster. Good news, you drill a discovery well.

Bad news, you drill dry holes north, east, south, and west of the discovery well. Worse news: you realize that your itty-bitty oil field will never pay for the four dry holes.

2 An old map in the geology library showed 30 wells scattered across Kansas that penetrated entirely through the sedimentary rock layers into the hard "basement" rock beneath. Another 40 wells reached the bottommost sedimentary layer without actually drilling into basement. The map was dated 1934.

Kansas is the most thoroughly explored oil province in the world. In a sense, someday everyplace will look like Kansas. I did take some criticism because Kansas is a political and not a geological entity. We later added a chunk of northern Oklahoma; our conclusions did not change.

The International Oil Scouts Association publishes annual lists of oil fields with dates of discovery, area of the field, depth, annual and cumulative production, and number of wells.[2] With the Kansas numbers in the computer, we could ask Menard's question in an expanded format. Of course, bigger oil fields tended to get discovered earlier, either by chance or by design. There is no possibility that a supergiant oil field lies undiscovered beneath Kansas today. There is not enough space in between the dry holes to fit in a supergiant. But how strongly does large size influence the order of discovery? This is more than an academic question. If the big oil fields in a province are already discovered, do we want to waste our effort plinking around finding the one-well fields?

Large size could increase the probability of being discovered in several ways. We start with the idea that each oil field has a characteristic length. "Length" is just a rough concept that stands for either the width, the thickness, or the length of the oil field. Here are examples of how "length" might be involved:

3 The *volume* of an oil field depends on the "length" cubed (length times width times thickness). Oil fields with big volumes might leak enough oil to the surface to attract a crowd of drillers.

2 The *area* depends on the "length" squared (length times width). This is essentially Menard's model. The bigger the area on the map, the more likely you are to hit it when throwing a dart.

1 The *diameter* is roughly the same as "length" (length raised to the power one). An example might arise if I start my exploration drill over what I think is the oil field, but if my location is wrong by more than the diameter of the oil field, it's dry hole time.

0 "Length" when raised to the zero power gives the number one. As an example, suppose that I go to Iran and drive in a surveyor's stake at the top of each of the visible surface anticlines. On top of each stake I mark a teensy dot, a mathematical point that has neither length nor width nor thickness; the dimension of the dot is the zero power of the "length." I put the names of the anticlines on slips of paper in a hat and draw names out of the hat in random order.

One of the Princeton statistics professors, Peter Bloomfield, wrote a one-line computer program that asked which exponent of the length (3 for volume, 2 for area, 1 for diameter, 0 for the location) best explained the Kansas history.[3] His answer was not constrained to be right on one of the integer numbers; it could (and did) fall in between. Bloomfield's best-fitting exponent was 0.66; discovery probability in Kansas fell between the zero and the first power of the "length." Bloomfield was my kind of statistician. On his office desk was a notebook labeled "CBS Election Night Manual"; he helped CBS "call" elections ahead of NBC or ABC: a statistician right on the firing line.

We then had the computer conduct "what if" searches for oil in Kansas. If exploration success depended on field area, the "length" squared, then the larger oil fields would have been discovered earlier than in real life, and some of the smallest oil fields would never have been discovered at all. This agrees with Menard's result. However, I put a different interpretation on the result. The raw Menard result would seem to say: "Boy, were we stupid. We could have done better by throwing darts at the map." My interpretation says that we were

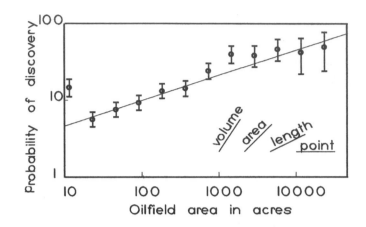

The probability of discovery increases with the size of an oil field, based on sizes of oil fields in Kansas discovered before 1977. However, the influence of size is much weaker than the area of the oil field. Discoverability is proportional to the field diameter raised to the 0.66 power, where "diameter" is the square root of the oil field area.

drilling for the good-looking targets and we were only slightly biased toward hitting the bigger targets first. Using the knowledge of 1930, East Texas was not a good-looking target.

I have been told that several Soviet exploration managers were sent to the far end of Siberia for not finding enough oil. Finally, a manager introduced a program of drilling exploration wells on a rectangular grid. All oil fields larger than the grid spacing were guaranteed to be discovered. Best of all, the exploration manager survived.

The Kansas data are incomplete, in the sense that some amount of oil remains to be discovered. Can we infer anything about the total "real" oil field population, including fields yet to be discovered? The answer is important in allocating future effort. Further exploration is worthwhile only where we can infer a substantial undiscovered population. Unfortunately, scholars are sharply divided into two camps, each camp publishing thinly disguised comments about the intelligence of their opponents. I'll try to be polite, but it isn't easy.

George Zipf, a mad-professor type at Harvard, noticed a statistical regularity in city populations and in word counts. He expanded his

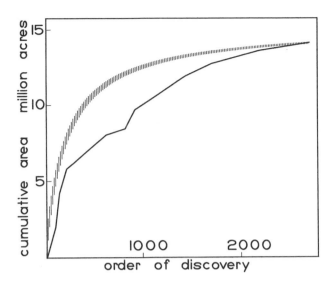

Repeated computer simulations of exploring Kansas by throwing random darts at a map fall in the dark band. Discovery probability would depend on the area, proportional to the second power of the diameter. The larger fields would have been discovered earlier than the actual history, shown by the solid black line.

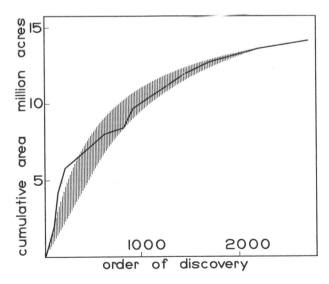

The dark band shows the range of multiple computer simulations of Kansas exploration using discovery proportional to the 0.66 power of the diameter. The simulations are a reasonably good match for the actual discoveries, shown by the solid line.

observations into a general natural law in a 1949 book.[4] Here is an example: the largest city in Belgium is Brussels. The second-largest city, Antwerp, is half the size of Brussels. Ghent, the third largest, is one-third the size of Brussels, and so on. Stated another way: multiplying the rank of the city times its population gives roughly the same number:

City	Rank	Population	Rank Times Population
Brussels	1	953,175	953,175
Antwerp	2	449,745	899,490
Ghent	3	224,545	673,635
Charleroi	4	203,853	815,412
• • •			
Lier	27	31,815	859,005

So let's try Zipf's law on the world's oil fields. A list of the 500 largest fields has been published, with help from the Petroconsultants data file.[5] The compilers, Sam Carmalt and Bill St. John, converted natural gas to equivalent amounts of oil. A barrel of oil is the energy equivalent of 6,000 cubic feet of gas. (The abbreviation *mcf* for 1,000 cubic feet appears on most household gas bills.) Here are the largest five, with samplings from further down the list:

Ghawar	$1 \times 87{,}500$	=	87,500
Burgan	$2 \times 87{,}083$	=	174,166
Urengoy	$3 \times 47{,}602$	=	142,806
Safaniya	$4 \times 38{,}066$	=	152,264
Bolivar	$5 \times 30{,}100$	=	150,500
Prudhoe Bay	$18 \times 13{,}783$	=	248,094
Troll	$36 \times 8{,}966$	=	322,766
East Texas	$56 \times 5{,}600$	=	313,600
Nar	$100 \times 2{,}333$	=	233,300
Koakoak	$200 \times 1{,}200$	=	240,000
Vat'yegan	500×500	=	250,000

Zipf's law is a little ragged for the 15 largest oil fields; the world's biggest oil fields aren't big enough. From oil field numbers 18 through 500, Zipf's law performs beautifully. Zipf's law seems to work for all but the largest oil and gas fields.

So what do we learn from applying Zipf's law? We learn that there is trouble at both ends, with big fields and with small fields. Let's go back to the populations of Belgian cities and ask whether there could be a city larger than Brussels. We could tour around Belgium searching for a larger city; looking for it might be as much fun as finding it. However, in present-day Belgium, we are not going to find a city with more people than Brussels. How about finding an oil field larger than the biggest known oil field? It is possible that a field bigger than Ghawar exists beneath the South China Sea. Even though I don't think it likely, I certainly have to admit that a larger field might exist. Zipf's law, in its simplest form, says nothing about replacing the leader.

Here's a way to be an optimist: suppose we discover two oil fields, both of them larger than Ghawar. Ghawar would be pushed down to third place, and the law would fit better. Does Zipf's law predict that two monster fields are yet to be discovered? Not exactly. On the other hand, the discoveries would please both SUV owners and statisticians. Since I don't know which countries will prevail in the dispute over the South China Sea, I will give arbitrary names to the first two imaginary discoveries:

"Imelda"	$1 \times 245{,}000$	=	245,000
"Ho Chi Minh"	$2 \times 115{,}000$	=	230,000
Ghawar	$3 \times 87{,}500$	=	262,500
Burgan	$4 \times 87{,}083$	=	348,332
Urengoy	$5 \times 47{,}602$	=	238,010
Safaniya	$6 \times 38{,}066$	=	228,393
Bolivar	$7 \times 30{,}100$	=	210,700
Prudhoe Bay	$20 \times 13{,}783$	=	275,660
Troll	$38 \times 8{,}966$	=	340,708
East Texas	$58 \times 5{,}600$	=	324,800
Nar	$102 \times 2{,}333$	=	237,966
Koakoak	$202 \times 1{,}200$	=	242,400
Vat'yegan	502×500	=	251,000

Zipf is now looking like a winner. George Zipf is not to be confused with George Gipp, the Notre Dame football player made famous by Ronald Reagan's movie role. We just won one for the Zipfer.

Zipf has a much more serious problem at the small end. In its simplest form, Zipf's law predicts that small oil fields, added together, contain an infinite amount of oil. Although there are fancy ways to do the calculation, it is simple if you get your computer to do the heavy lifting. Here is a five-line program that will run in Qbasic, available on early versions of Microsoft Windows. (Bill Gates started on his road to riches by selling a version of Basic that he hadn't yet written. A free version of Qbasic can be downloaded from www.winsite.com.) The comments after the single quotation marks do not have to be included.

```
sum = 509.2 'billions of barrels in the largest 17 oil fields
FOR n = 18 to 10000 'for fields numbered 18 to 10,000
    sum = sum + 233.3 / n '233.3 / n is the Zipf-law size of the field
NEXT n
PRINT sum, 233.3/ n 'total oil, size of smallest field
```

Run the program with the number of fields set to 10,000, then 100,000, then 1 million. With each run, the sum gets bigger by 537 billion barrels, and the smallest field becomes a factor of 10 smaller. Current world production is 24 billion barrels of oil per year. Zipf's law seems to be telling us that we can extend the oil supply by 22 years simply by utilizing oil fields a tenth as large as our present fields. This is a conclusion only an economist could love.

Zipf's law can't go on forever: the smallest oil field has to be larger than one oil molecule. A more practical limit is the smallest oil field that pays back the cost of exploration and production. A few years ago, exploration geologists for a major oil company were happily finding new oil fields in the Williston Basin, in North Dakota and Montana. They were sharply criticized by their management because they were finding only a million barrels at a time. To most of us, a million barrels at $30 per barrel is more than pocket change. Almost 50 years ago, a senior geologist with Standard of New Jersey (now ExxonMobil) realized that the company's profitability depended entirely on the giant oil fields.[6] The conclusion applies to consumers as well as to oil companies: half the world's oil is contained in the 100 largest fields. The search is for elephants, not squirrels. Today, there are variations on Zipf's law.[7] Some versions behave like a playground teeter-totter: if you squeeze one end to avoid having unlimited oil in little oil fields, the other end predicts ridiculous numbers of supergiant fields.

There is an alternative way of looking at natural size distributions, with roots older than Zipf's law. Do the following experiment: collect a bag of natural sand from the beach or from a river; the source doesn't matter. Sort the sand grains according to size by sifting the sand down through a stack of wire-mesh screens (obviously, with the coarsest screens at the top of the stack). Weigh the amount of sand retained on each screen and make a graph of the results. The graph looks lopsided on ordinary, linear graph paper. However, geologists learned, as early as 1914, to use logarithmic graph paper, which shapes the sand weights into a nice, symmetrical bell-shaped curve.[8] The lines on plain graph paper are numbered 1, 2, 3 . . . with equal spacing between the numbers. On logarithmic, or "log," paper the numbers 1, 10, 100, 1,000 are equally spaced from one another. The bell-shaped curve on an ordinary linear graph is called a "normal" curve; if it has a symmetrical bell shape on log paper, it is called a "lognormal" curve.

The normal curve arises when a cluster of independent numbers get added together. The normal curve is also known as a "Gaussian" curve, because the shape came from a study by "the prince of mathematicians" Karl Friedrich Gauss.[9] Here's another bitty Qbasic program that generates an approximately normal curve.

```
SCREEN 12 'opens a graphics screen
DIM sum(600) 'reserves 600 memory spaces
FOR n = 1 to 50,000
        k = 100 * (RND + RND + RND + RND + RND + RND)
        sum(k) = sum(k) + 1
        PSET (k, 400 – sum(k)), 15 'plots a point
NEXT n
```

Each RND command generates a random number between 0 and 1. The middle of the random number range is 0.5, so adding up six of them will usually give a number around 3. Rarely, but sometimes, all six numbers will be small, and adding them up will give a number slightly above 0. Similarly, if all six are close to 1, the sum will be just under 6. The whole collection makes a bell-shaped curve with its center at 3 and its tails extending toward 0 and 6. The normal curve is supposed to happen if we all run the 100-yard dash. There are a few speed merchants, most people are in the middle, and I'm puffing along in the back. Or listen to popcorn pop: a few early birds, fast popping in the middle, a few stragglers at the end.

The normal curve comes from *adding* together several numbers that arise independently. Deffeyes' rule, mentioned in an earlier chapter, says that if anything goes wrong, you get a dry hole. You need a good source rock, the source rock has to visit the oil window, the reservoir rock must have porosity and permeability, the cap rock better not leak, and there has to be a trap geometry. Instead of adding up the score, in this game the scores on each part are *multiplied* together. If any one of the scores is 0, you get a 0 for the whole ball game. Here's the payoff: multiplying numbers together is the same as adding up the logarithms of the numbers. Adding up independent numbers gets you a normal curve; multiplying the same numbers together generates a lognormal curve.[10]

Back to the sand grains: suppose that instead of weighing the grains on each screen, I had *counted* the grains. (Not me, actually. The purpose of summers is to provide student summer hires for really dull tasks.) In the finer sizes, there are a lot more grains for the same weight. The sand grains, in all sizes, are usually made of the mineral quartz, which has a uniform density. The number of grains of one size depends on the cube of the weight. Something really nice happens: changing from weighing grains to counting grains is only a matter of renumbering the scale at the bottom of the graph. The lognormal curve retains its original size and shape.

So how do the world's oil fields look on a lognormal graph? Not bad, but they did not look bad on the Zipf plot either. Usually, you cannot decide which fits best simply by eyeballing the graphs. At least the lognormal distribution does not blow up at either end. The mathematical form of the lognormal does allow the possibility that an ex-

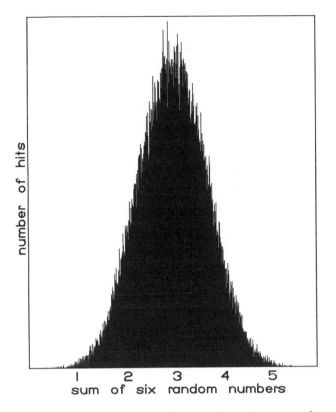

Adding together six independent random numbers gives a rough approximation to a Gaussian bell-shaped curve.

ceedingly large undiscovered oil field might exist. However, the lognormal curve assigns an extremely low probability to the existence of that super-duper-giant. The expected amount of oil in the population of hypothetical monsters (size multiplied by lognormal probability) is small. The message from the lognormal curve is, "Don't get your hopes up."

Neither Zipf's law, and its variations, nor the lognormal curves give an entirely satisfactory picture of the world's oil fields, at least not satisfactory enough to dictate a strategy for future energy supplies. Does that mean that no relevant knowledge exists? Both approaches explain that oil fields have a similarity to automobile accidents: there are many little fender benders but very few multicar flaming pileups on the freeway. However, liability lawyers know to chase the bigger accidents. I have a silly worry that illustrates the problem. Official statistics report that the scheduled airlines are far safer, per passenger mile, than travel by automobile. That is reassuring. However, I don't really care about passenger miles; how about the fatality rate per vehicle mile? Do the 300 people packed in with me on an Airbus multiply by 300 my chances of getting killed?

The conclusion for future oil supplies is that oil fields bigger than 2 billion barrels contain more than half of the existing oil. I expect that an extremely intensive program of future drilling would, at most, lower that midpoint to 1 billion barrels. If there is a future oil supply, it has to be anchored by big oil fields.

Because major oil fields have a statistical resemblance to serious automobile accidents, giant fields are irregularly distributed around the world. (Golda Meier: "Don't talk to me about Moses. He spent 40 years leading us to the one place in the Middle East that doesn't have any oil.") Here are the discovery dates for the first billion-barrel field in major petroleum regions:

South America	1868	Brea, Peru
Czarist Russia	1870	Surakhanoskoye, Azerbaijan
North America	1871	Bradford, Pennsylvania
Middle East	1908	Masijid-I-Suleman, Iran
Southeast Asia	1929	Seria, Brunei
China	1938	Lochunmiao, Gansu
North Africa	1956	Hassi Messaoud, Algeria
North Sea	1969	Ekofisk, Norway

Of course, there can be disagreement about what constitutes "discovery." The general message is clear: major oil provinces were identified fairly early, but some pleasant late surprises came along later in North Africa and the North Sea.

The list of early discoveries emphasizes that oil production was an international phenomenon from the beginning. In the town of Miaoli, on Taiwan, I was pleasantly surprised to see a life-size sculpture: roughnecks tonging drill pipe. Miaoli belongs alongside Bakersfield, California; Aberdeen, Scotland; Morgan City, Louisiana; and

Baku, Azerbaijan. Americans have had a huge, but not overwhelming, presence in the oil industry. In Rotterdam Harbor, the spot market buys and sells oil in U.S. dollars for a 42-gallon barrel.

You could tell that the Middle East would be a major oil province by reading the Bible: a bush that was burning but was not consumed, a pillar of cloud by day and of fire by night, the fiery furnace in the Book of Daniel. In 1927, a well drilled half a mile from the fiery furnace discovered the Kirkuk field: 17 billion barrels.[11] The Middle East arrived gradually. You can entertain either or both of these hypotheses:

1 There was little economic incentive to develop Middle East fields because oil prices stayed below $3 per barrel until 1970.

2 There was a conspiracy among the major oil companies to keep additional oil off the market.

Here are dates of the first billion-barrel discovery in each Middle East country:

Iran	1908
Iraq	1927
Kuwait	1938
Saudi Arabia	1938
Abu Dhabi	1954

Really large production from Saudi Arabia began after the 1948 discovery of Ghawar, the world's largest known oil field at 87 billion barrels. The Arabian (Persian) Gulf consists of continental crust, bent down slightly below sea level. Oil production extends right across the gulf. The Arabian side of the gulf includes a number of small, oil-rich countries, sometimes grouped as the United Arab Emirates.

The political history of the Middle East, including the effects of oil, is an oft-told tale. Whole books have been written about the history;[12] I'm not about to rewrite them. Here is a short list of historic events relevant to future oil supplies:

1 Aramco, the operating oil company in Saudi Arabia, had a drill-or-drop provision in its original contract. If oil had not been dis-

covered on portions of the original concession area, drilling rights to those areas would revert to Saudi Arabia: use it or lose it. With that motivation, Aramco discovered a string of multi-billion-barrel oil fields. In stages from 1976 to 1980, Saudi Arabia nationalized Aramco, but the company as an entity continued.[13] In contrast, in Iraq and Iran, nationalization was accomplished by ejecting the former consortium that operated oil exploration and production.

2 Oman is a country at the far eastern end of the Arabian peninsula. I would characterize Oman as being on the distant thin fringe of the Middle East oil province. Earlier Middle East exploration arrangements usually consisted of an agreement with an international consortium of oil companies. Oman allowed some of the midsize "independent" oil companies to participate in exploration. All the oil discoveries in Oman are smaller than a billion barrels, but about 30 oil fields have been discovered. The oil field map of Oman oil fields is beginning to look like an oil map of central Oklahoma. It makes me wonder whether the entire Middle East is loaded with small oil fields.

3 Iraq and Iran are the big puzzles. Despite their similar names, the two countries have very different histories and cultures. In Iran, 24 oil fields larger than a billion barrels were found from 1908 to 1974. In Iraq, 11 fields larger than a billion barrels were discovered between 1927 and 1979. Despite the smaller number of fields in Iraq, both countries have roughly 100 billion barrels of oil reserves in the existing fields. Political events, and the Iraq-Iran war from 1980 to 1988, shut down further petroleum exploration. (As I read about the Iraq-Iran war at the time, I had a ghastly fantasy that the United States was helping whichever side was behind, the purpose being to trash both countries.) Which country would discover the most oil if serious exploration, using advanced techniques, were to resume? I had been saying that it was Iran, because Iranian oil fields are located on spectacularly visible surface anticlines. A retired petroleum geologist told me, "Nope, it's Iraq. We plugged and abandoned any

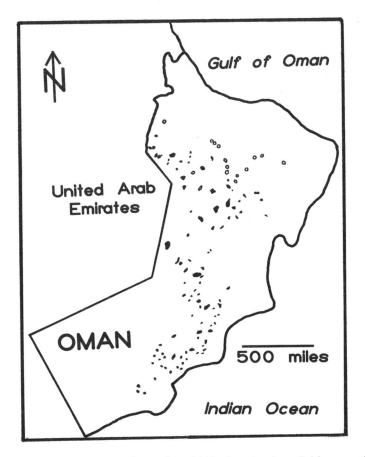

Known oil fields in Oman are shown in solid black; natural gas fields are outlined. The large number of small fields discovered in Oman hints that numerous un-discovered small fields may exist elsewhere in the Middle East.

well that wouldn't make 5,000 barrels a day. Threw 'em back in the water."

Whether it is Iraq or Iran, the undiscovered oil in the Middle East is very likely the largest untapped supply in the world. I am not invent-ing spectacular political scenarios to restart Middle East oil explo-ration. Very likely, future oil production is going to be more valuable as a petrochemical feedstock than it is for fuel. The remaining political barriers in Iraq and Iran may save petrochemical raw materials for future generations.

The North Sea is the most recent major oil and gas province to be discovered. There is a preamble to the discovery. When the Germans overran the Netherlands in 1941, many Shell employees were able to leave. (What we all call "Shell" is actually Royal Dutch/Shell. The Hague manages exploration and production; London runs the finances.) The Gestapo held a gun to the head of the highest-ranking remaining Shell employee and demanded that he find oil. During World War II, the Germans were desperate for oil; they were synthesizing aviation gasoline from soft brown coal. The Shell man drilled an unbroken string of dry holes, but he looked so professional doing it that the Nazis never pulled the trigger. After the war, they brought in the giant Groningen field in the Netherlands. Guaranteed way to make your Gestapo agent feel bad.

The Groningen gas field extended from land out into the North Sea. The Ekofisk discovery in 1969 established that major amounts of oil existed out in the middle of the North Sea.[14] The geographic centerline of the North Sea and the centerline of the oil fields coincided almost exactly; Norway and Great Britain produce almost identical amounts of oil.

The oil and gas had an enormous effect on the British economy. In years preceding the North Sea discovery, the humor magazine *Punch* was offering advice to the English about adapting to third-world status. The economy of Great Britain in 1967 was in bad shape and going downhill. Today, Great Britain is the world's tenth-largest producer of oil; Norway is sixth.[15] Those two countries not only escaped the shock of importing oil at high prices, but they exported high-priced oil. The British tradition of miserably cold homes in winter was repealed by natural gas from the North Sea.

It did occur to me that the centerline rule would assign most of the British share to Scotland instead of to England. I drew in the midpoint line on a map: 90 percent of the oil spoke with a Scottish burr, 10 percent had an Oxbridge accent. Virtually all the population of Scotland already knew that. Without Scottish oil, England really *would*

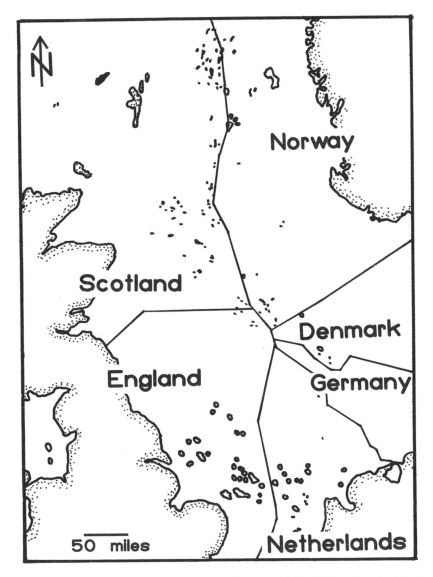

My unofficial homemade map shows that the bulk of the British oil in the North Sea would belong to Scotland and only a small portion would go to England. Oil fields are shown in solid black, and gas field are outlined.

be a third-world country. (On field trips to Scotland, I told my undergraduates that the Scots became a conquered people at about the same time that the United States gained its independence: Culloden, 1746; Yorktown, 1781.) The English are willing for Scotland to enjoy a few symbols of independence as long as the Scots don't get uppity about their oil.

The situation in Norway was rather different. Norwegians gained their independence from Sweden and Denmark in 1905. Norway had a small population, but not a particularly wealthy population. With the oil bonanza, Norwegians took a comparatively long-term view. Not the kind of people to indulge in an immediate spending orgy.

Both exploration and production in the North Sea have been on a fast track. The North Sea province will have a lifetime, from discovery to depletion, about half as long as those of other major petroleum provinces. There are two reasons:

1 The reflection seismic method works well at sea; discoveries were made very rapidly during the 1970s. Great Britain had a strong economic need for speedy discoveries and rapid production. The government promptly put up offshore areas for companies to bid for drilling rights.

2 Most of the North Sea consists of one family of drilling targets. Once that family was understood, success after success followed in rapid succession. Other petroleum provinces, particularly geographically large provinces, usually contain several different types of oil fields that were discovered at widely separated times. In that sense, west Texas is a composite of several different "North Seas," not all developed at the same time or discovered by the same methods.

In the center of the North Sea, there are no geographic names to borrow for oil field names. A few have names like Block 16/26, but most discoverers had more imagination. On the Norwegian side are Troll, Gullfaks, and Frigg. Among the many names on the British side are oil fields named for four early British geologists: Hutton, Murchison, Miller, and Lyell. Thank you, Conoco.

Mid-Kansas-Transcontinental Yates #A-30 sending a 10-inch stream of oil 200 feet high from a depth of 1,070 feet. The official proration test of this well on September 23, 1929, gave it an hourly production of 8,528.4 barrels of oil, or a daily potential of 204,681 barrels—the largest production actually gauged for any well. AAPG © 1929; reprinted with permission of the AAPG, whose permission is required for further use.

I have long been curious about the world record; the most productive single oil well. The standard answer is the Potrero #4 in the Golden Lane of Mexico, but the quoted figure of 200,000 barrels per day is probably too large. Confusion arises for two reasons: (1) Most of the early candidates blew "wild" for days to months before the well was brought under control. Therefore, the reservoir was partially depleted before a production test was possible. (2) The usual measure of size was to flow the well wide open into storage tanks for 24 hours. My father insisted that the Mary Sudik #1 at Oklahoma City, the legendary

"Wild Mary," never had a proper 24-hour test. The well filled all the tanks in Oklahoma City in 12 hours.

We are left with an "honor roll" of wells larger than 100,000 barrels per day. Notice that any one of these wells would generate a cash flow larger than the gross national product of some United Nations countries.

1910	Portrero del Llano #4, Veracruz, Mexico, wild for 60 days
1910	Aguila #4, Veracruz, Mexico, wild for 60 days
1912	Lakeview #1, Midway field, California, 90,000 bbl/day
1928	Mary Sudik #1, Oklahoma City, wild for 11 days
1929	Yates #A-30, west Texas
1948	Masjid-i-Sulaiman, #7-7, Iran
1956	Alborz #5, Iran, wild for 90 days

CHAPTER 7

Hubbert Revisited

Even if M. King Hubbert had said nothing about future oil production, he would deserve an important place in the history of geology. His analyses of groundwater flow, size scaling, Darcy's law, hydrodynamic oil traps, and great thrust faults were huge contributions. Hubbert's 1956 warning was not something to be dismissed lightly.[1]

Although journalists occasionally mention "scientific authorities," there are no authorities in science. Scientists can, and do, make mistakes, even after receiving honorary degrees, high government appointments, and Nobel Prizes. However, when Linus Pauling gave his 1955 opinion that atmospheric nuclear weapons tests were a health hazard,[2] a lot of us paid attention. Pauling's long, successful previous track record gave a statistical expectation, if not proof, that he would be right about nuclear tests. (When Pauling was elderly, he was not able to prove his hypothesis that large doses of vitamin C would cure the common cold. However, just in case, I drink a big glass of orange juice every morning.) Similarly, Hubbert's track record suggested to the rest of us that he might, just might, be right about the future of the oil industry.

Hubbert grew up in the central Texas town of San Saba. He was named Marion King after a teacher that his parents admired. Hubbert had a slightly asymmetrical face from a boyhood accident, when a log rolled over him. He found his way to the University of Chicago, which was then the most significant innovation in American education. (Some say it still is.) In the late 1930s Hubbert spent seven years with the rank of instructor at Columbia University without ever being promoted to assistant professor, an experience he recounted frequently and bitterly. He spent the next 25 years with Shell; after reaching Shell's retirement age, he joined the U.S. Geological Survey (USGS) and taught occasionally at Stanford.[3] When I asked Hubbert later how his teaching went, he replied that it went well. A Stanford student's reply to the same question was that the students were terrified.

Hubbert's later association with the USGS is relevant to the larger history. The USGS is not your typical bureaucracy. For more than 100 years, it set high standards for detailed field mapping, laboratory work, and carefully edited publications. Sometimes the USGS was a little slow, but its work was of the highest quality. However, when USGS workers tried to estimate resources, they acted, well, like bureaucrats. Repeatedly, they used statistically dubious estimation methods. For instance, at one point they divided the United States into little areas and asked geologists to guess how much oil was under each area. The USGS then added up the guesses *as if the areas were independent* and got overly optimistic answers. Unfortunately, small adjacent areas are not independent; if no source rock was present or if the rocks had been buried deeper than the oil window, then a whole bunch of little areas are eliminated all at once. Hubbert's presence at the USGS from 1964 to 1976 did not cure the tendency; in 2000 the USGS again released implausibly large estimates of world oil.[4]

I mentioned in chapter 1 that neither Shell nor the rest of the oil industry was ready to listen to Hubbert's prediction. It was as if a physician had diagnosed virulent, metastasized cancer; denial was one of the responses. In April 2000, the president of the American Association of Petroleum Geologists published yet another "Hubbert was wrong" opinion. I think most people have an initial "take"—to accept or to deny Hubbert's conclusion. The ensuing arguments are largely attempts to justify the original intuitive opinion. Put another way, I have never known anyone to switch sides as a result of an intellectual analysis of Hubbert's methods.

A favorite form of the anti-Hubbert argument is to expand Hubbert's original analysis of U.S. oil production to include more geography (offshore and Alaska) or an untapped oil source (Colorado oil shale). Here is my favorite way of proving that Hubbert was wrong: we invite the Saudis to join us as our 51st state. Make them an offer they can't refuse. (Congratulations. You get to elect two United States senators and these here gemmen are from the Internal Revenue Service.) Now you count the Saudi oil as U.S. oil; Hubbert is toast. Obviously, first you draw a line around the problem and then discuss what goes on inside the line. Fudging the line, either geographically or categorically, only makes a mess out of the discussion.

The numerical methods that Hubbert used to make his prediction are not crystal clear. Today, 44 years later, my guess is that Hubbert, like everyone else, reached his conclusion first and then searched for raw data and methods to support his conclusion. (Despite sharing roughly 100 lunches and several long discussions with Hubbert, I never had the guts to cross-examine him about the earliest roots of his prediction. Lunch discussions were more cheerful when Hubbert chose the topic.) Guessing the answer first and then searching for supporting arguments is a common scientific procedure; it is not cheating. Hubbert had a message; he packaged his message in a format that he found convincing.

Why, then, should we be relying on a messy 44-year-old method to predict an upcoming world oil shortage? The standard folklore: Hubbert's prediction, made in 1956, that U.S. oil production would peak around 1972 was right on the button. The actual peak year was 1970. In fact, Hubbert gave his readers a choice between two estimates. The leaner estimate predicted a peak in 1965, and his more generous curve gave his celebrated prediction.

Let's begin with an illustration, although it was not one used by Hubbert. The amount of anthracite coal mined each year in Pennsylvania makes a beautiful bell-shaped curve.[5] The only stray bulge in the curve is a spurt in production during World War II. So what explains the bell shape? From 1830 to 1900, anthracite production grew by 5 percent each year. Growth by a fixed annual percentage is a mathe-

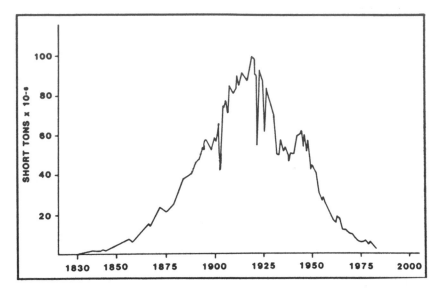

The annual production of anthracite coal from Pennsylvania displays a bell-shaped curve. Major deviations from the curve are a dip during the Great Depression (1929–38) and a peak during World War II. AAPG © 1986; reprinted with permission of the AAPG, whose permission is required for further use.

matical process that expands without limit, variously called exponential growth or compound-interest growth. After 1900, the finite supply of anthracite coal begins to pinch; the peak production year is around 1920. The decline, almost to zero, does not mean that every last ton of anthracite has been mined and sold. Thicker and more accessible coal beds get mined first. Thin and steeply dipping coal beds are difficult and dangerous to mine. The decline in production has two causes: the easily mined beds are gone, and the difficult beds are not competitive in price with other fuels.

Could anything prevent the production history from being bell-shaped? Glad you asked.

1 Production could be terminated abruptly because some horrible toxic effect was found to be associated with the product. Amphibole asbestos from South Africa is an example.
2 Production of anthracite coal in Pennsylvania could halt suddenly because an exploration geologist found an anthracite bed

on the other side of the world that was thick, of high quality, and right next to a seaport. The 1970s discovery of huge, high-grade deposits of uranium in northern Australia and Canada almost closed down all other uranium mines.

3 The top of the curve could be chopped off because the market was saturated. You can sell only a certain amount of Tabasco sauce, no matter how low the price. In the 1930s, Texas and Oklahoma could saturate the oil markets. Since 1970, Saudi Arabia has enjoyed a production capacity that could saturate the world market.

The anthracite example shows that a bell-shaped history can happen. Skeptics may wish to recall the story about a mathematician, a statistician, and a logician riding the train from London to Edinburgh. As the train passed into Scotland, the mathematician saw a black sheep on the hillside and said, "Sheep in Scotland are black." The statistician corrected him, "At least one sheep in Scotland is black." The logician chimed in, "At least one sheep in Scotland is black, on one side." At least anthracite production had a bell-shaped history, in one state. It happens.

Let's try an example from the oil fields. For the moment, we define the target as the area on land in the lower-48 United States. We look at the history of finding oil by recording the dates on which each oil field was discovered.[6] Hubbert used a different definition of "discovery," so I'll call these "hits." Bringing in the first producing well in an oil field is the date of the "hit." It may take years, sometimes decades, of drilling to measure the full extent of the oil field. On the other hand, nobody is going to "undiscover" it. Once production is established, it is inevitable that the entire field will eventually be drilled up. You don't "hit" it a second time.

The lower-48 "hits" history contains profoundly bad news. More oil was found in the United States during the 1930s than in any decade before or since.

1 I'm 69 years old; before I finished grade school, almost all the good stuff had been discovered. The last billion-barrel hit on

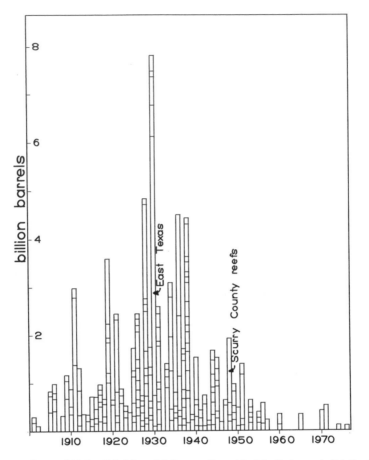

Discovery dates of U.S. oil fields exhibit a scattered but bell-shaped distribution. All oil fields larger than 100 million barrels on land in the lower 48 states are plotted against the year of the first successful well in the field. Despite the Great Depression, more oil was found in the decade from 1930 to 1940 than in any decade before or since. Notable large fields are East Texas (1930) and the Scurry County reefs (1948).

land in the lower 48 was the Scurry County, Texas, reefs in 1948. I was still in high school.

2 A pop quiz for economists who insist that higher oil prices bring on discoveries: What did we call the 1930s? (the Great Depression) What was the price of oil? ($1 a barrel; without production rationing the free-market price would have been 10 cents a barrel)

This cloud has a tarnished silver lining. The hits history has its good years and its bad years, but it does look like a bell-shaped curve.

Hubbert's model of a production history started off with two assumptions, assumptions that can be confirmed only if they fit the actual history.

1 The history is bell-shaped, with a rounded top and tails on both ends.

2 The decline side of the curve is a mirror image of the initial increase.

The Pennsylvania anthracite discussion suggests possible reasons why the decline might not mirror the increase: an environmental restriction, the appearance of a competitor with lower production costs, or market saturation. In choosing a mirror-symmetrical curve, Hubbert was following one of the oldest traditions in science: try the simplest explanation first. The concept is usually attributed to William of Occam, who formulated it before the year 1350. A preference for the simplest explanation is sometimes called "Occam's razor." So, for now, until proven otherwise, let's keep the simple idea of a mirror-symmetrical bell-shaped curve.

If looking at the oil production history, by itself, would solve the problem, you wouldn't need me. You wouldn't need Hubbert. We'll borrow one final bit of magic from Hubbert's methods. Before you can produce oil, you have to have it identified: corralled and hog-tied. As explained in chapter 5, producible amounts of oil and gas in the ground are reported as "reserves." Reserve estimates allowed Hubbert, and allow us, a look into the future. Of course, there will be some new discoveries, but most of the oil that will be produced during the next 10 years is already identified in today's reserves.

Before taking on data for the whole world, let's begin with the United States. The purpose is to gain some perspective by looking at a mature area. Validating the method in a well-explored area gives us some confidence when approaching the less mature world picture. Describing the United States as "mature" does not mean that all the oil has already been produced. The United States is currently the world's second-largest oil-producing country. (Saudi Arabia is the largest.) "Maturity" is shown by 563,160 actively producing oil wells in the United States, versus 1,560 in Saudi Arabia.

Hubbert actually used two different methods, each appropriate for a different level of maturity:

1 Fourteen years before U.S. production reached its peak, Hubbert used two educated guesses about the total amount of conventional oil available.[7] He showed what a bell-shaped curve of crude oil production would look like when constrained by one of the guesses.

2 Eight years before the production peak, Hubbert added estimates of oil reserves (still in the ground) to the cumulative oil already produced to estimate the eventual total oil production.[8] He no longer had to rely on the two educated guesses.

First, we will rerun both of Hubbert's strategies using U.S. data up to the year 2000. We are not asking whether "Hubbert was right." Neither Alaska nor the offshore Gulf Coast was producing major amounts of oil at the time of Hubbert's initial estimates.

When we start to "fit" a mathematical curve to a set of numerical observations, there are several choices. Hubbert initially did not justify his choice of methods; we are not certain he was even aware of all the available alternatives. Hubbert published one "fit," and his implied message was, "Here it is, take it or leave it." We can exercise a wider freedom. Here are the choices:

1 There are three different bell-shaped curves in common use. The Gaussian (also called "normal") curve has a blunter top and narrower tails on each side. The Lorentzian (also called Cauchy) has a narrower top and wider tails. The logistic curve is in between. Hubbert used the logistic curve in all his analyses.

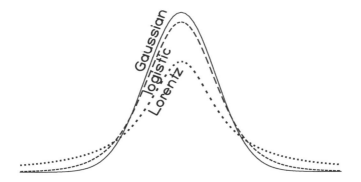

The three commonly used bell-shaped distributions, plotted with the same total area under the curve and with the same width at half the maximum height. The Gaussian distribution (solid line) is slightly wider near the peak but is much narrower on the flanks.

2 We want to minimize the misfit between the observed data and the mathematical curve. However, there are two different meanings of "minimize" in statistics. Here is an example: you have graded your students' exams and want to calculate a class average. The usual procedure is to add up the grades and divide by the number of students. However, consider the effect of the lowest grade. Was the lowest score a failing grade, like 65 percent? Or did the student turn in a blank exam paper for a zero? In a class of a dozen or so students, the difference between 65 and 0 moves the class average down by about 6 percent. An alternative is to report the "median" grade: half of the students got a higher grade than the median, and half of the students got a lower grade. The student in the middle got the median grade. (It helps to have an odd number of students.) The important point: the median grade does not depend on whether the lowest grade is a 65 or 0. As applied to more complex problems, the "average" minimizes the sum of the *squares* of the misfits between the observations and the mathematical model. The median minimizes the sum of the misfits, but with all the misfits made positive (the sum of the "absolute values" of the misfits). The least-squares method pays great attention to the larger misfits. The least absolute value technique tries to get close to most of the observations and pays little attention to the large misfits.[9] If a few of the observations have been written down wrong (called "blunders"), you don't want to be using a least-squares fit.

3 You can make a best fit to the numbers, as ordinary numbers. However, there are reasons for transforming the numbers before starting. For instance, with stock market prices you are interested in the *percentage* changes; it makes sense to take the logarithm of the prices. The graphs on stockmaster.com have a logarithmic price scale. If you have counts of events instead of amounts, it helps to use the square root of the counts. (The classic example of counts is the annual number of deaths from horse kicks in the Prussian cavalry.) Statisticians refer to these tactics as "fitting in the logarithmic domain" or "fitting in the square root domain."

Which is going to work best for the oil numbers? We are dealing with amounts of oil, and we want to focus on the years of large production; therefore, we have no reason to go to the logarithmic domain or to the square root domain. We can eliminate blunders by plotting a graph of the oil production against time; if the points on the graph follow along smoothly, then there are no major errors in getting the numbers into the computer. On the oil data, least squares and least absolute value fits give essentially identical answers. I use the least absolute value in everything that follows.

The big choice is among the Gaussian, logistic, and Lorentzian curves. By far the best data set for making the choice is the U.S. history of oil production, including Alaskan and offshore production. The Gaussian fits like a glove, the logistic fits less well, and the Lorentzian misses badly. So we choose the Gaussian for the rest of this chapter. Notice that Hubbert originally had only the left-hand half of this curve available in 1956; there was no way he could have tested whether his choice of the logistic curve was the best.

Does the best-fitting Gaussian curve predict the future? There are two kinds of uncertainty, one kind inside the statistical model and another kind outside. Inside the model, there is a simple way of handling

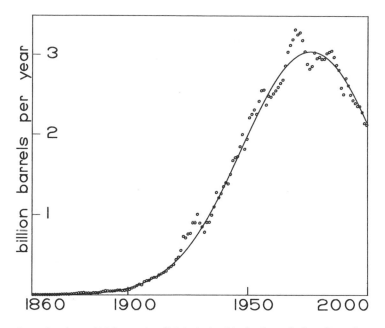

Annual production of U.S. crude oil (circles) with the best-fitting Gaussian curve superimposed as a solid line. Production from Alaska and from offshore oil fields is included.

the problem. The "best" fit is the Gaussian bell-shaped curve that min-imizes the misfits between the mathematical curve and the observed oil production. In addition to the "best" fit, I want to know the range of answers that are "pretty near" the best. I choose a number that is 1 percent larger than the sum of the errors in the "best" model and display all the "pretty good" results.

Here I have a confession to make: I much prefer running these statistical fits using what the computer scientists dismiss as "brute force." The Gaussian curve has three available adjustments: the year of peak oil production, the amount of oil produced during that peak year, and the number of years between the half-maximum points. I simply write three loops inside one another in the computer program. The outermost loop tries out different peak years, the intermediate loop tries out values for the peak production, and the innermost loop tries different widths. It takes my middle-of-the-line desktop computer a few seconds to a few minutes to try all the possibilities. There are more sophisticated programs that would find the "best" model in milliseconds; I much prefer to be able to mess around with the problem on my own terms.

The range of "pretty good" models gives me a feel for the internal uncertainty in the estimate. Is there anything *outside* the estimate that could make a drastic difference: severe worldwide economic recession, disastrous climate change, thermonuclear war? Well, they might, but notice that the existing U.S. production curve already includes the effects of the Great Depression, two world wars, and Jimmy Carter.

The great weakness of Hubbert's 1956 prediction was his reliance on educated estimates of the eventual total oil production. In 1962, he improved on his original work by sweating the eventual total U.S. oil production out of the historical data.[10] Hubbert noticed that oil reserves were beginning to drop even though production was still rising. He carried out the analysis using cumulative curves. For instance, cumulative production is all the oil produced in the United States from the 1859 start, up to a given year. Hubbert defined cumulative "discovery" to be the cumulative production up to a given time *plus* the underground reserves known at that time. (The reason for using cumulated curves is their smoothness. I recently tried doing the same analysis without cumulating. The year-to-year annual numbers jitter around so much that the analysis looked hopeless. Actually, it is the same analysis; the cumulative curves make a better visual impression.)

With the advantage of hindsight, we established earlier in this chapter that the Gaussian curve fits the U.S. production history better than the logistic curve that Hubbert used. So, in updating Hubbert's analysis of U.S. oil production, we will use the Gaussian. There are two data streams, one of which is the cumulative production history known from 1859 through to the end of 1999. The second is cumulative discovery: cumulative production plus reserves. Reserve estimates are available from 1946 through 1999. Cumulative production history will be matched to a cumulated Gaussian curve (called an "error func-

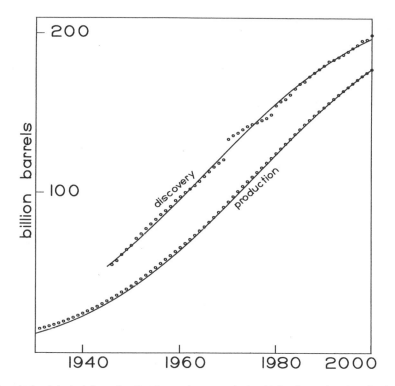

The circles labeled "production" are the cumulative U.S. oil production (including Alaskan and offshore oil) added up from 1860 to a given year. The "discovery" circles are the cumulative production up to a given year *plus* the reserves reported for that year. The jump in the discovery curve in 1970 is the addition of the Prudhoe Bay field in Alaska to the reserves. The solid lines are the best-fitting cumulative Gaussian curves, with the constraint that the discovery and production curves have exactly the same shape but are displaced in time. The best fit occurs with discoveries leading production by 11 years.

tion" in the trade). The cumulative discovery history will be matched by another cumulated Gaussian, but the mathematical production curve and the discovery curve are constrained to have identical shapes; the only difference is that the two curves are displaced from each other by a constant number of years. This constant displacement in time is a piece of Hubbert's heavy magic. It is another example of trying the simplest hypothesis first. The computer is allowed to fiddle around trying different spacings between the discovery curve and the production curve, but the computer is not allowed to choose one spacing for the early history and a different spacing for later times. One size fits all. For the U.S. history, the best fit puts the discovery curve 11 years ahead of the production curve. The discovery curve is a predictor of future production; it was this trick that allowed Hubbert to see into the future.

Now for the hard part. It's time to take on the world oil history. The first step is to repeat Hubbert's 1956 use of educated guesses about the total world oil; the second step is using the discovery curve to estimate the total.

World oil production history is available for the years from 1850 through 2000.[11] All the produced oil goes to refineries; there is probably a reasonably accurate production inventory. The largest uncertainty is Soviet oil production during the Cold War; there is neither a dip nor a bulge in the historical curve suggesting a serious misrepresentation. In order to use Hubbert's original 1956 strategy, we need a couple of educated guesses about the total world oil that will eventually be produced from conventional oil wells. Colin J. Campbell, working with the Petroconsultants data files, has made country-by-country estimates, and the sum of his estimates is 1.8 trillion barrels.[12] We find the Gaussian curve that best fits the world production history *and* has 1.8 trillion barrels under the total curve. That curve reaches its peak production in the year 2003. This is one version of a hard-core Hubbertian prediction: world oil production will start to fall after 2003.

A few oil experts believe, and all economists know, that Campbell's 1.8 trillion is too small.[13] For some years, I have been using 2 trillion as a rough number with no claim to precision. In one of his last papers, Hubbert used 2 trillion as his probable guess.[14] Although a few exceedingly large estimates have been suggested, a reasonably generous upper guess is 2.1 trillion barrels. Fitting the world production history to a 2.1-trillion-barrel Gaussian gives peak production at the end of the year 2009. I emphasize that this is a "what if" calculation. The 2.1 trillion number was a test of how far we might postpone the peak production year.

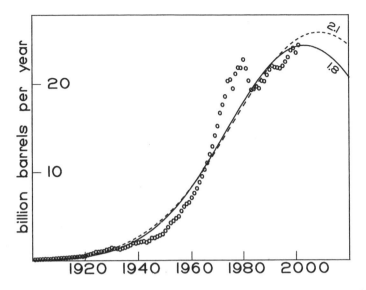

Annual production of world oil (circles), with Gaussian curves corresponding to total eventual oil recovery of 1.8 and 2.1 trillion barrels. A steeper rising curve with its top chopped off by market limitations would make a better fit.

Hubbert's second method used the reserve estimates to peek into the future. Chapter 5 mentioned that the reported reserves for some OPEC nations increased abruptly during the 1980s. If the increases are sudden, we can see them and exercise some judgment about their effect. For instance, Iraq's reserves were reported at an even 100 billion barrels for the years 1985 through 1998. During those years no new wells were drilled and oil was being produced, but the reported reserves never changed. A reserve estimate of exactly 100 billion doesn't *prove* that it is only a rough guess. (This year, I repeatedly drove the same business trip; several times I punched the trip odometer as I left my garage. The round trip was exactly 100 miles; I'll have to tell the IRS that it was either 99 miles or 101 miles.) However, producing oil with no new discoveries has to be lowering the reserves. In order to get more reasonable reserve estimates, I subtracted out any abrupt jump during the 1980s from each OPEC country's reserves. After the 1980s, I estimated that 60 percent of the production was a drawdown

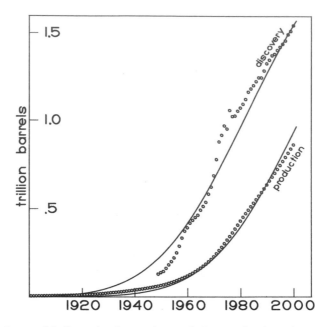

Cumulative world oil production and cumulative production plus reserves (discoveries) are compared with the best-fitting cumulative Gaussian curves. The two solid curves are constrained to have exactly the same shape but are displaced in time. No estimates are fed into the computer program that fits the curves to the data. The best-fitting estimate is for the year of maximum production to be 2003, and the eventual cumulative production is estimated at 2.12 trillion barrels. Discoveries lead production by 21 years.

from the reserves and 40 percent of the production was either corrections for previous underestimates or the addition of new oil reservoirs. The 60–40 split is intended as an average performance figure for those OPEC countries that reported abrupt reserve increases. In Iraq, there is little or no drilling activity that could be replacing reserves. In contrast, Saudi Aramco is pursuing an intensive program: bringing in new fields and drilling horizontal wells to improve recovery from existing fields.

Now we can use Hubbert's second method: fitting parallel curves to the cumulative production and to discoveries (cumulative production plus reserves). No educated guesses go in. The constraints are the

Gaussian shape of the history and a constant spacing between the cumulative production and the discovery curves.

The resulting estimate gives a peak production year of 2003 and a total eventual oil recovery of 2.12 trillion barrels. The peak year, 2003, is the same year that we got by fitting Campbell's 1.8-trillion-barrel estimate to the production history. Other published estimates, using variations on Hubbert's methods, give peak years from 2004 to 2009. I honestly do not have an opinion as to the exact date for two reasons: (1) the revisions of OPEC reserves may or may not reflect reality; (2) OPEC production capacities are closely guarded secrets. If your country has surplus production capacity today, you are A Player in the global oil game. If your wells are currently producing to capacity, you are merely a spectator.

This much is certain: no initiative put in place starting today can have a substantial effect on the peak production year. No Caspian Sea exploration, no drilling in the South China Sea, no SUV replacements, no renewable energy projects can be brought on at a sufficient rate to avoid a bidding war for the remaining oil. At least, let's hope that the war is waged with cash instead of with nuclear warheads.

CHAPTER 8

Rate Plots

There is a fascinating parallel between studies of population growth and future oil predictions. The population studies began more than 100 years before Hubbert's first analysis of U.S. oil production. In 1798, the Reverend Thomas Malthus wrote that any population that grows by a fixed percentage each year eventually outgrows its food supply.[1] The analogy is to a compound-interest savings account, in which the interest paid in is a fixed percentage of the money currently in the account. Up through 1955, there was a widespread assumption that the U.S. oil industry would always be able to keep up with a consumption increase of 5 percent per year. In 1956, Hubbert pointed out that it would require an infinite amount of oil to keep up with a fixed-percentage annual increase.

In 1838, Charles Darwin read what Malthus wrote and began thinking about evolution.[2] In the same year, Pierre Verhulst in Belgium suggested a modification of Malthus's unlimited-growth theory. Verhulst introduced the idea of a "carrying capacity," the maximum population that could be sustained by the environment.[3] He then suggested that the population growth rate would depend on the fraction of the carrying capacity that was not yet occupied, and in 1982 Hubbert used the same equation for the fraction of undiscovered oil.[4] When the population is small compared with the carrying capacity, population growth takes the same compound-interest form suggested by Malthus. Later, the growth slows down to zero as the population approaches the carrying capacity. Verhulst named the graph of population versus time a "logistic" curve. In modern English usage *logistic* makes us think of arranging supplies for an army, but an earlier meaning was associated with numerical calculations.

In his original 1956 paper, Hubbert used a logistic curve to describe U.S. oil production.[5] Most of us at the time were puzzled, because we presumed all bell-shaped curves to be Gaussian until proven otherwise. In a long paper published in 1982, Hubbert finally justified his use of the logistic curve: it had an appealing symmetry in the underlying mathematics. In his 1982 paper, Hubbert refers to the original Verhulst papers and goes through a complete derivation of the mathematics. It is not clear whether Hubbert was aware of an entire century of work carried out by population biologists.

Although the mathematical equations are identical, the analogy between population growth and oil production seems a little odd. Oil wells don't have babies. Some pieces have obvious counterparts.[6] The carrying capacity for a population is similar to the total amount of oil that can be produced from the ground. For oil, it makes good sense that the probability of discovering additional oil at any moment depends on the fraction of the total oil that still remains undiscovered. The peculiar part is the analogy between people having babies and oil wells begetting additional oil wells. In a crude sense, oil wells do raise families. Drilling a discovery well brings on a bunch of new wells to develop the oil field. Understanding the geology of a new oil field leads to a search for similar fields.

In his 1982 paper, Hubbert developed a clever way of making a graph that turns a logistic curve into a straight line. However, neither Hubbert nor later workers seem to have used the technique to plot oil data. Population biologists had long been using the method, including real data.[7] Their horizontal axis was simply population; their vertical axis was the "per capita growth rate." For oil, the horizontal axis is the cumulative amount of oil produced. The vertical axis is the new oil produced per year as a percentage of the cumulative production up to that year.

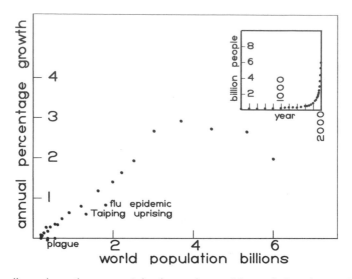

The small graph on the upper right shows the world population during the past 2,000 years. Although the abrupt rise is astonishing, the eye picks up very little detail from the small graph. The larger graph plots annual percentage population growth on the vertical scale and the population on the horizontal scale. (The Taiping uprising was an 1860 religious crusade in China with 50 million casualties.) As we will see later, world population is not following a logistic curve.

This growth-rate-versus-production graph can be used equally well to plot discoveries. In Hubbert's definition, *discoveries* are the cumulative production plus the known reserves for a given year. The new oil found in any year is the cumulative discovery as of the end of the year minus the cumulative discovery at the end of the previous year. The graph then consists of cumulative discovery on the horizontal axis, and the vertical axis is the new oil found as a percentage of the cumulative. As explained in earlier chapters, reserve estimates have considerable uncertainty. In fact, Hubbert's "discoveries" will go backward whenever the downsizing of reserves exceeds the annual production.

The best part is plotting both production and discovery data on the same graph. When the wells finally run dry, the cumulative production has to equal the cumulative discoveries. If the discovery points are scattered on both sides of the production points, then the

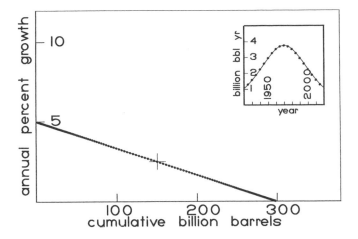

On this idealized graph, the cumulative amount of oil produced is plotted on the horizontal scale. The vertical scale is the annual production as a percentage of the cumulative amount produced up to that year. On this graph, a logistic bell-shaped curve plots as a straight line. The dots are one year apart and fall closer together at the beginning, and at the end, when only small amounts of oil are being produced. The intersection of the straight line with the horizontal axis gives the total amount of oil that will be produced when the oil fields in that region are finally exhausted. The intersection with the vertical axis determines the width of the bell-shaped curve. The cross at the center marks the peak year of annual production, when half of the oil has been produced. The small graph inset at the upper right shows the same curve in the more familiar plot of the annual production for each year.

discovery and production histories have the same shape but are displaced by a constant time. Best of all, discoveries are a guide to likely future production. If the history is approximately a straight line on the graph, then the history is well described by a logistic curve.

But what if the bell-shaped curve is actually Gaussian instead of logistic? The same graph still works reasonably well. Since the Gaussian curve has narrower tails, the ends of the curve are scrunched in a bit toward the middle. However, about two-thirds of the oil is produced in the middle, where the Gaussian curve is reasonably straight. Readers with calculus skills are encouraged either (1) to find a similar procedure, with linear axes, that straightens out a Gaussian[8] or (2) to prove that no such procedure exists.

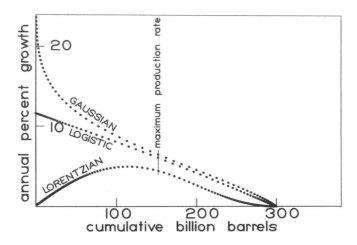

On a growth rate versus cumulative graph, a Gaussian curve lies reasonably close to the straight logistic line through the latter part of its history. However, on the left the Gaussian curve is well above the logistic. The U.S. and the world oil production histories have the same feature, suggesting that at least their initial years are close to Gaussian. The Lorentzian history has a bump in the center, which does not match any of the actual production histories examined to date.

We turn first to the United States, not through patriotism but because of the long production history and prompt annual revisions of reserve estimates. The open circles for discoveries scatter nicely on both sides of the solid circles for production. The single discovery circle above the general trend represents 1970, when the 9-billion-barrel Prudhoe Bay field in Alaska was entered into the U.S. reserve estimates all in one year.

For the U.S. history, a reasonably good straight line starts at 5.5 percent and goes through the rightmost solid black dot for the year 2000 production. Extending this line to the horizontal axis gives a final ultimate production, when the last dog is dead, of 220 billion barrels. This is not far from the 200 billion that gave Hubbert's 1956 best prediction, or from the 210 billion estimated by Campbell in 1997.[9] The best-fitting Gaussian curve for the U.S. production and reserves in chapter 7 gave an expected ultimate production of 220 billion barrels. Unfortunately, the 362-billion-barrel estimate issued by the U.S. Geo-

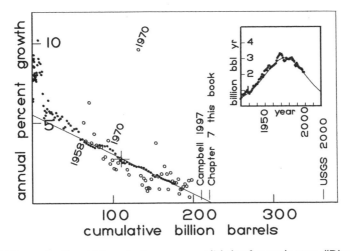

The U.S. production history is shown as a solid dot for each year. "Discoveries" (cumulative production plus reserves) are plotted as open circles. The high discovery point for 1970 comes because the 9-billion-barrel Prudhoe Bay field in Alaska was added to the U.S. reserves in a single year. The best-fitting line starts on the left at 5.5 percent and ends at 220 billion barrels. The peak year of the logistic curve is 1975, although the actual year of maximum production was 1970. Other estimates of the total recoverable U.S. oil are shown at the bottom of the large graph. The recent estimate of 362 billion barrels made by the U.S. Geological Survey seems implausibly high.

logical Survey in 2000 is way out in right field.[10] To make the USGS estimate come true, there would have to be new U.S. oil discoveries that add up to the reserves of Kuwait.

The "fast track" depletion of the North Sea deposits, mentioned in a previous chapter, can be quantified by plotting the production and reserves data for Norway. Instead of the U.S. rate of 5.5 percent, the Norwegian data start at a 16 percent rate. Stepping along the best-fitting line, using the spacing between the recent annual steps, would suggest that the maximum Norwegian production is only a year or two away. The discovery data (open circles) are so noisy that they do not do much to constrain the estimate of future production. At least, the discoveries do not reject the placement of a line toward 30 billion barrels of ultimate production.

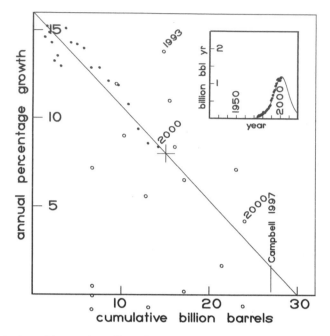

Solid dots show Norwegian oil production, all from the North Sea, as an example of fast-track depletion. The straight line hits the vertical axis at 16 percent, in contrast to 5.5 percent for the United States. Open circles mark discoveries. The four leftmost open circles reflect the reserves shown by the initial exploration wells, before production began. The most recent production dot is for the year 2000. It looks as if one more year will carry Norway to the peak production year shown by the center cross. Examining only the solid dots on the small graph would lead any of us to think that the annual production could continue to rise for several more years.

For world oil, the production history lurches around before it settles down in 1983 to a straight line. The left-hand part of the production history lies above the line because the curve is closer to Gaussian than to a logistic distribution. In addition, there is a local valley in 1942 and a local peak in 1970, which I will leave for economic historians to explain.

The straight line for world oil begins with annual production at 5 percent of cumulative production and ends at 2 trillion barrels. Lots of us, Hubbert included, have used 2 trillion for years because it was

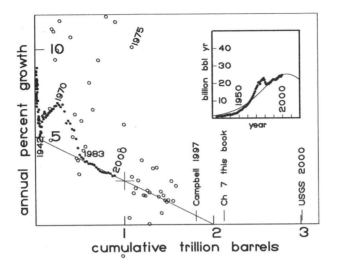

World oil production (solid circles) and discoveries (open circles) are consistent with a straight-line trend after 1983. The line starts at 5 percent and ends with 2 trillion barrels of production. Warning: I have removed abrupt increases in reserves announced by some OPEC countries during the 1980s. Starting from the most recently completed production year (2000), it will take about five years to reach the mathematical peak.

a nice even number. The discovery circles certainly encourage extending the line exactly to 2 trillion. The USGS estimate is again implausibly high. Its number, 3.012 trillion, requires discovering an additional amount of oil equivalent to the entire Middle East.

On the world plot, the serious bad news is the low level of exploration success after 1979. (Remember, however, that I have edited the reserves numbers after 1985 to take out abrupt increases in reserve reports by some of the OPEC countries. Campbell did similar, but not identical, editing in his 1997 treatment.)[11] There are recent success stories, particularly along the west coast of Africa, but nothing that compares to discovering the major North Sea oil fields during the early 1970s.

So when does world oil production peak and start downward? That's the big enchilada. You can use the spacing between the recent production dots and see that four or five more dots will carry us to the plus sign that marks the midpoint. Once we draw that straight line through the year 2000 dot, the logistic curve is fully defined. The mathematical peak falls at the year 2004.7; call it 2005. However, I'm not betting the farm that the actual year is 2005 and not 2003 or 2006. The top of the mathematical distribution is smoothly curved, and there is a fair amount of jitter in the year-to-year production. Remember, the center of the best-fit U.S. curve was 1975 and the actual single peak year was 1970. There is nothing plausible that could postpone the peak until 2009. Get used to it.

CHAPTER 9

The Future of Fossil Fuels

Years ago, a visitor to Switzerland asked how people earned a living in those high Alpine villages. "Each village gets paid for doing laundry for the next village." Today, an equally silly answer would be, "Each village uses the Internet to sell toilet paper to the next village." The lesson for the global village: we can't all work in the service economy; somebody has to be down grubbing at the base of the economic pyramid. The list of fundamental activities is short: agriculture, ranching, forestry, fisheries, mining, and petroleum.

Consider the names on some great art museums: Getty (oil), Guggenheim (copper), de Menil (oil services), Gulbenkian (oil). Huge wealth used to be accumulated at the base of the economic system. Today's dot.com billionaires up at the apex of the pyramid may not be able to see all the way down to the base. They might have the illusion that everyone can earn a living by selling software to the next village.

A permanent drop in oil production will pull one of the blocks out from underneath the pyramid. The previous chapter strongly suggests that the drop will happen in this decade. Major disruptions likely will follow. What should we do? The question exists at two levels:

1 What can individuals and institutions do, in their enlightened self-interest, to minimize the impact of a global oil shortage?

2 As a society, how can we rearrange the global economy to lessen our dependence on oil?

Republicans choose line 1; Democrats pick line 2. The division is not that simple. I'm a registered Democrat, but I still feel authorized to protect myself while the world gets its act together. That's why line 1 says "*enlightened* self-interest." This chapter discusses fossil fuels; chapter 10 treats alternative energy sources.

A "fossil" is the remains of an ancient organism. A fossil fuel is solar energy stored by organisms in ancient times. A major lesson: the source of the world's oil accumulated over hundreds of millions of years; most of the world's oil has been discovered during my lifetime. Large-scale use of fossil fuels began with the Industrial Revolution, of which coal was an integral part. In a sense, the fossil fuels are a one-time gift that lifted us up from subsistence agriculture and eventually should lead us to a future based on renewable resources.

A fad of 10 years ago was "finding oil on Wall Street." Publicly traded companies with undervalued oil and gas reserves became targets for stock traders and merger-and-acquisition specialists. An individual today could offset his or her family's petroleum consumption by purchasing oil and gas company shares. But "petroleum consumption" is not just gasoline and home heating. Many other goods and services include an energy cost. Only comparatively wealthy individuals have the flexibility to offset future oil price rises by owning oil stocks. Many of us in the middle class have the majority of our assets tied up in home equity and managed retirement funds; we have little flexibility to change investments.

My friends would love to have a short list of trading symbols for companies with undervalued U.S. oil reserves. Unfortunately, professional traders have their computers monitoring these opportunities. Several large oil companies are ready to swallow smaller companies to acquire additional reserves. I'm not likely to find anything in a three-month-old copy of *Oil and Gas Journal* that the full-time traders have overlooked.

Colleges and universities are substantial energy users; the more fortunate schools have sizable endowments. Investing a block of the endowment, directly or indirectly, in oil and gas reserves could be a prudent choice. I can report, however, that I got absolutely nowhere when I tried to explain that to Princeton University in 1980. Two universities have their energy problem solved in advance: the University of Texas at Austin and Texas A&M share revenues from state-owned oil lands.[1] The University of Texas at Austin has an endowment second only to Harvard; an oil price rise increases UT Austin's income.

Here is a second limitation: we can't all buy Texaco. For years, Texaco has been a Wall Street favorite: large oil and gas reserves, few employees, low costs. Texaco is disappearing as an investment opportunity; it is being acquired by Chevron. In fact, the entire stockholder equity for all U.S. petroleum companies combined is not large enough for all of us to offset the threatened oil price rise. Equities are, at best, an opportunity for smart money to get in during the early rounds.

Gradually, the role of the major oil companies has changed. They are becoming more like enormous service companies, although they do bring along their own investment capital. The majors appear to be vertically integrated companies, active in everything from exploration to marketing. However, production, transportation, refining, and marketing are almost independent activities. The oil has a defined price in between each stage. I was disappointed when I learned that the gasoline sold in a Shell station might come from a Conoco refinery. Shell writes specifications and analyzes the gasoline to see that it meets the specifications, but another company may do the actual refining. There is no equivalent of a chateau-bottled wine.

In the past 10 years, major oil companies have been getting as little as 10 percent of the oil as their share in new overseas ventures.[2] Much of the oil revenue goes to the producing nation, usually through the participation of government-controlled oil companies. At the marketing end, a substantial part of the price of gasoline at the pump is another round of taxes. Let's face it, oil used to be a highly profitable business, and governments figured out how to cut themselves in for a big piece of the action. As world oil production decreases, some governments may feel more pain than the major oil companies.

Years ago, the New York Times Company bought enough Canadian forestland to grow pulpwood as fast as it prints newspapers. Heavy petroleum users among large corporations, like FedEx and UPS, could attempt a takeover of an oil company large enough to offset their consumption. In the past, a useful way of insuring major producers and consumers against the effect of a price changes was purchasing futures contracts. However, the ordinary futures contracts extend for a year or two. The oil problem extends for 10 years or more. Anyone who agrees to supply oil 10 years from now, for a price agreed on today, very likely will disappear into bankruptcy before the contract matures.

The financial world has reorganized since 1980. Effects of oil price rises, during the late 1970s, took months to years to spread from industry to industry; from price increases to wage demands. In the new economy, the shock of an oil price rise will spread in milliseconds. Detailed computer models of the world economy will flash messages to buy or sell everything from AT&T to Xerox, from aluminum to zinc, from bahts to zlotys.

Oil and gas companies have split personalities. Finding a new oil field is an investment that generates income for 20 years, but Wall Street has a fixation on the most recent quarterly earnings. The stock market has preempted the time horizon for most corporations, especially for those outside the natural resource industries. The CEO is assigned stock options on a two-year time scale. His motivation is to get the stock price up in two years: one way is to close the research lab, fire the engineers designing the upcoming product, make the quarterly profits look great, retire and cash out. Of course, I have overstated the case, but the top managers should not be given heavy personal incentives to ignore long-term goals.

A second struggle about the time horizon comes from interest rates. A potential project, in any industry, is judged by the present-day discounted value of the expected future cash flow.[3] At times of high inflation, and high interest rates, the future barely exists. During the 1980 oil crisis, interest rates reached 20 percent. At a 20 percent discount rate, a dollar earned 10 years in the future has a present-day value of only 11 cents. Long-term efforts, like designing a new aircraft or exploring for new oil, become difficult to justify. ExxonMobil is not designed to be a charity. In the existing system, ExxonMobil is not asked to take on uneconomic projects to provide us with future oil. Discounting the future is more than a problem for the oil industry. Economists and the Sierra Club have opposing philosophies.

The dichotomy between the short and long time scales affects employees, including the professional staff. Oil companies have to survive boom-and-bust cycles. This one circulated around Denver in 1970:

Q: How do you address a petroleum geologist?
A: Waiter!

Usually at a time of low prices, all the companies in the industry are hurting simultaneously. Of the 200 largest U.S. oil and gas companies, 133 had a net loss for the year 1998. Employees who are let go rarely find employment at another oil company; many of them leave the industry and never want to return.

At the other extreme, around 1980 our geology undergraduates were starting out at salaries higher than those of their professors. Some were very capable, but some were not our best students. Boom years can be just as distorted as lean years. However, falling oil production is not necessarily bad news for individual petroleum geologists. My rough calculation shows that geologists' salaries amount to about two cents for each barrel of U.S. oil. There's plenty of room to raise that to four cents per barrel.

In 1997, I looked into direct investment in wells. Could I set up a company to buy and operate wells? During 1998, inflation-corrected oil prices approached the levels of the 1930s Depression. At the same time, a terrible drought was affecting the midcontinent. Rumor had it that the Oklahoma City police chief reported the good news that only

13 women in Oklahoma City were trying to earn a living as prostitutes. His bad news was that 7 of them were still virgins. It seemed like a good time to be shopping for oil wells.

I picked out one category of wells that might be made marginally profitable, but my major motivation was holding the wells for five years until world oil production peaked. The search focused down to certain counties and a particular depth range. (Don't ask! Just because I am writing a book doesn't mean I have to tell *everything*.) The potential investors I approached had two reactions: (1) oil was a severely depressed industry, and (2) you could make a gazillion dollars in 1997 by buying Internet stocks. In the year 2000, potential investors told me that oil prices had recently doubled and oil wells would be too expensive to buy. So it goes.

In the long run, the eventual use for oil will be for manufacturing useful organic chemicals. I expect our grandchildren to ask, "You burned it? All those lovely organic molecules, you just burned it?" Sorry, we burned it. Originally, the feedstock for synthesizing organic chemicals was a tar by-product from treating coal. The change from coal tar to oil, roughly from 1930 to 1950, gave the name "petrochemical" to the industry. Today, about 7 percent of world oil production goes into petrochemical manufacturing.[4]

The modern state of Israel is, in one limited sense, a product of the synthetic chemical industry. During World War I, a shortage of the solvent acetone was severely limiting British production of gunpowder. Chaim Weizmann, a young chemical engineer working in England, developed a successful method for synthesizing acetone.[5] After the war, as a thank-you gift for Weizmann, the British carved out an independent country in the Middle East, which they named Palestine.

Natural crude oil consists of relatively stable molecules. No surprise, over geologic time all the unstable molecules break down. Early petroleum refineries simply sorted out those stable molecules into salable products. Oil production, and oil refineries, had been running for 50 years before the automobile was invented. Originally, their most marketable product was kerosene, primarily for light from kerosene lanterns. The Oklahoma City oil field was discovered by the Indian

Territory Illuminating Oil Company, founded before Oklahoma became a state. *Oil for the Lamps of China* was a novel and a movie of the 1930s.[6]

Introducing the automobile generated a huge market for gasoline, but the amount of gasoline that could be separated from crude oil was low, 10 or 20 percent. The original separation process was distillation, sorting out molecules that boiled at different temperatures. Later, commerce copied nature; thermal cracking was introduced to make smaller molecules, salable as gasoline, out of larger molecules.[7] The route to petrochemicals was opened because thermal cracking produced some unstable molecules, molecules that would participate in further chemical reactions. Before World War II, both American and German companies developed commercial petrochemical products. After the war, petrochemical production expanded enormously, in both the range of products and the tonnages produced. A whole range of plastics and synthetic fibers became part of our daily lives. (An unusual application: to clean a lens, use the freshly broken surface of a white foam-plastic packing "peanut." The plastic is made from oil and gas; there is no grit in it.)

A major early petrochemical complex grew up around the Houston Ship Channel. Large petroleum refineries supplied hydrocarbons. Salt from the salt domes was turned into hydrochloric acid, chlorine, and sodium hydroxide. Sulfur, originally from the caps of salt domes, was a source for sulfuric acid. Bromine came from seawater. Portions of the complex were owned and operated by various companies; a local maze of pipelines conveyed intermediate products from one plant to another.

We don't usually think of oil processing as a source of fertilizer. However, nitrogen compounds are important in plant nutrition; the first of the three numbers on a fertilizer bag gives the nitrogen content. Mineral sources of nitrogen compounds are quite rare. The obvious source is the atmosphere; air is 76 percent nitrogen. In 1908, Fritz Haber reacted atmospheric nitrogen with hydrogen gas to produce ammonia; other nitrogen compounds could in turn be produced from the ammonia.[8] Today, the cheapest source of hydrogen gas is hydrocarbons: oil and gas. Virtually all our nitrogen compounds are indirectly petrochemicals.

Major oil companies, such as ExxonMobil, are vertically integrated; they explore for oil, drill wells, transport oil, run refineries, and their gasoline hose reaches all the way to your car's fuel tank. On the side, major oil companies are also petrochemical producers. Profits, or losses, happen all along the chain from exploration to marketing. Oil-producing countries soon realized that crude oil was leaving their borders and refining and petrochemical profits were appearing elsewhere. Gradually, oil-producing countries arranged for the building of local refineries and petrochemical plants. Today, petrochemical plants in Saudi Arabia use about 10 percent of the country's oil production.[9]

As the global economy eventually converts to renewable energy, oil and gas will continue to be produced for petrochemicals, lubricants, and specialty products. An example of a specialty product: "Vaseline," a trade name for petroleum jelly, is made by Chesebrough-Ponds. One of the companies split up in 1911 from the original Rockefeller-owned Standard Oil was Chesebrough, along with Exxon, Mobil, Chevron, Arco, and Amoco.

If there is a continuing use for oil, what can we do to extend the supply? More oil can be squeezed out of existing fields, oil can be produced from tar sands and oil shales, and natural gas production can expand.

Conventional primary and secondary recovery methods typically leave half of the oil still in the ground. A depleted oil field can be acquired cheaply; lots of clever people have devised methods for recovering the second half of the oil. In 1980, when oil reached $37 per barrel, projects known as "tertiary recovery" or "enhanced recovery" were initiated. By 1986, 512 projects were operating in the United States.[10] Tertiary recovery looked like the wave of the future. Since 1986, however, the total amount of U.S. oil produced from these enhanced methods has been roughly constant at 700,000 barrels per day, compared with U.S. production of 6 million barrels per day. The number of operating projects dropped from 512 (in 1986) to 199 (in 1998).[11]

Rather than listing all the tertiary recovery methods, real and imaginary, here are the most successful methods in terms of U.S. oil production in 1998:

1 Steam injection, either into separate injection wells or as "huff-and-puff" intermittent injection into production wells: 420,000 barrels per day. Steam is the oldest of the enhanced methods; it is used to recover thick, viscous oil, largely in California.
2 Carbon dioxide injection to increase the volume of the oil and to reduce its viscosity: 179,000 barrels per day.
3 Hydrocarbon injection: 102,000 barrels per day.
4 Eleven other methods combined: 58,000 barrels per day.

Of course, the lack of growth was partially explained by the 1997 drop in crude oil prices. If there are major price rises associated with a drop in conventional oil production, we can expect that these methods will be dusted off and expanded once again. However, the 1980s experience shows that it is not duck-soup easy to pull out the second half of the oil.

Of the unconventional sources of oil, recovery from tar sands is being expanded rapidly. A tar sand is essentially a dead oil field. When erosion brings an oil field to the surface, the smaller molecules evaporate, and a nearly solid tar is left in the reservoir rock. Although tar sands exist around the world, Alberta contains two enormous tar sand deposits: Athabaska and Cold Lake. Tarry oil is extracted by mining the sand, contacting it with hot water, and separating the oil. To tell the truth, the oil isn't all that great; it contains a lot of sulfur. The original tar sand recovery plant, opened in 1978, was profitable in the sense that oil sales paid the operating costs, but the plant took forever to pay back the capital investment.[12] Improved methods followed; as of 1999, 2 billion dollars is being invested in new tar sand operations in Alberta. Closely related to tar sands are reservoir rocks filled with "heavy" oil, oil that is too viscous to move using ordinary production practices.

If a tar sand is a moribund oil field, an oil shale is an unborn oil field. An "oil shale" contains neither oil nor shale; it is an ordinary pe-

In the Athabaska tar sands of Alberta, huge trucks are used to haul tar-impregnated sand from open pit mines to the processing plant. Economies of scale are very important in making a previously marginal activity profitable. © Jonathan Blair/CORBIS.

troleum source rock that has never been buried into the oil window. A particularly large oil shale deposit exists where Utah, Colorado, and Wyoming come together. As the first transcontinental railroad reached the town of Green River, Wyoming, a work crew gathered up a circle of rocks to surround their campfire. The rocks caught fire. They were not coal; the heat from the campfire caused thermal cracking to produce oil, and it was the oil that was burning. The rock unit, the source rock, was eventually named the Green River Formation. After the formation was deposited, mountain ranges arose that broke the original lake basin into several pieces. The Wyoming portion is the Green River Basin, in Utah it is the Uinta Basin, and in Colorado it is the Piceance Basin. ("Piceance" is a local name, made to look French again by early map makers. The word is pronounced "piss ants.") In the western Uinta Basin, some of the source rocks have been buried

The Green River "oil shale" is an oil source rock deposited in a nonmarine lake. The bottom water in the lake was depleted in oxygen, organic matter was preserved, and there were no bottom-dwelling animals to churn up the thin layers of sediment.

into the oil window; two midsize oil fields produce oil whose source is the Green River Formation.

The Green River Formation turned out to be unusual in several ways. It was not marine; it was formed in a saline lake. Almost half of the world's supply of sodium carbonate is mined from the Green River Formation.[13] Unique minerals occur in the Green River Formation. Spectacular fossils occur because no scavengers could live on the oxygen-free lake bottom. And, for our purposes, the oil that could be released from the Green River oil shale is roughly equal to all the world's conventional oil.

When I was an undergraduate, crude oil sold for $2.50 per barrel, and everyone expected that if oil got to $5.00 per barrel, oil shale would come on the market and put conventional oil fields out of business. However, each time the price of oil increased, the imaginary oil shale price was always about $3 per barrel above the current price. I've been waiting for 50 years; what's wrong? Several pilot plants have operated, mostly in western Colorado. The rock has to be mined, crushed, and heated in closed containers. The leftovers after the oil is recovered fluff up to more than their original volume; the hole where the rock was mined isn't big enough to hold the waste. Several variations on the theme were developed, but none of them seemed economic at $25 per barrel. Research continues.

Up until 1953, natural gas sold for less than 10 cents per 1,000 cubic feet. One barrel of oil has the same energy content as 6,000 cubic feet of gas. In 1953, oil sold for $2.70 per barrel; 6,000 cubic feet of gas cost 60 cents. If you were a consumer, natural gas was a bargain. If you were drilling for oil, gas was a nuisance. Gradually, the school of hard knocks identified areas that produced only natural gas as places to stay away from. Usually, these were areas that had been buried beneath the oil window at some time in their history. Natural gas prices went above $2 in 1981. Blow the dust off the old maps; there's a new ball game in town! Places that were now too deep, or had once been too deep, were ripe for drilling. In contrast to oil, future supplies of natural gas can be expanded by additional exploration and deeper wells.[14]

Natural gas has some advantages over the other fossil fuels. Of the fossil fuels, natural gas adds the smallest amount of carbon to the atmosphere. In natural gas there are four hydrogen atoms for each carbon; gasoline and diesel fuel average two hydrogen atoms per carbon, and coal can be almost pure carbon. When burned, hydrogen produces water, which for the moment has not been listed as a harmful substance. Sulfur removal from natural gas is cheaper and more complete than sulfur removal from oil or coal. As a fuel for house heating and for power generation, natural gas causes the least environmental damage of the fossil fuels. Not zero damage, the *least* damage.

The big advantage of gasoline is portability; the gasoline tank does not occupy much space or weight in an automobile. A tank of gasoline will take the car 300 to 400 miles. The only way to make natural gas portable is to compress it to high pressure in thick-walled tanks. Even at high pressure, a car with half the trunk filled with natural gas tanks will travel only 150 miles between fuel stops. Com-

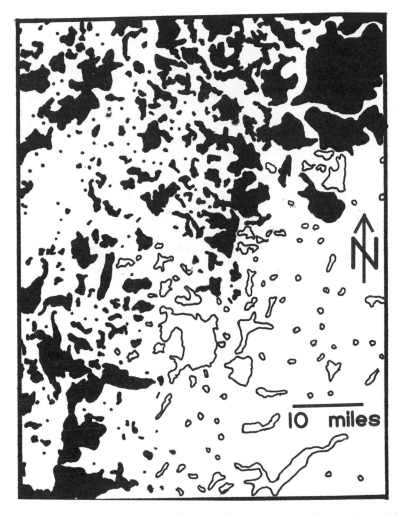

Sometimes on an oil and gas map, it is possible to locate the former base of the oil window at the line where oil and gas production changes to all gas. On this map of a portion of southeastern Oklahoma, oil fields are shown in solid black, and natural gas fields are outlined.

muters, school buses, and construction vehicles typically travel much less than 150 miles per day.

The eye opener is this: natural gas costs 57 cents for the equivalent of a gallon of gasoline. That's based on the retail price of natural gas delivered recently to my basement for the furnace, stove, and hot

High-pressure tanks being installed to convert a conventional pickup truck to run on compressed natural gas. A typical configuration consists of two tanks at the front of the pickup bed. Additional tanks are in the background. © Sergio Dorantes/CORBIS.

water heater. At a meeting on natural gas, one of the exhibits was a compressor previously used to fill scuba divers' air tanks, modified to compress natural gas. We all crowded around that exhibit, fantasizing about driving around on cheap fuel. The government hasn't figured out how to tax natural gas as a motor fuel, which made the fantasy even sweeter. In Italy and New Zealand, roadside natural gas filling stations have been around for 20 years.

Of course, I wondered about safety. What happens if I have my trunk half filled with compressed natural gas tanks and I am in a rear-end collision? I was told that the gas tanks are so strong that the car will crumple around the gas tanks like so much aluminum foil. Also, a gas flame burns upward. The ghastly hazard with gasoline is the fuel pouring down on the ground, then burning upward on the people.

Except for the portability problem, natural gas is a wonderful automotive fuel. The efficiency of an ordinary automobile engine increases with the compression ratio: how much the fuel-air mixture can be compressed before the spark plug initiates burning. However, burning can begin spontaneously and prematurely (called "pinging") if chemical bonds between carbon atoms in the fuel start to break. The compression ratio that a fuel will tolerate is the "octane" number displayed on a filling station pump. The scale is based on two standard molecules: isooctane taken as 100 on the scale and heptane taken as 0. Regular gasoline is about 85 octane, premium gasoline is 90 octane, and top grades of aviation gasoline are 100 octane. Natural gas is 135 octane. An automotive engine designed from the beginning to run on natural gas can have a high compression ratio and a high efficiency. Of course, we all respond to our gut feelings more than to numerical arguments: in the hotel lobby at a petroleum engineers meeting a year ago was a full-size, mean-looking drag racer that ran on natural gas.

Diesel engines avoid the pinging problem by compressing only air, then squirting in the fuel to initiate burning. There are hybrid diesels that compress a mixture of natural gas and air and then squirt in a little diesel fuel to act as a spark plug. I thought the hybrid diesel was a really clever new idea. I was told that the hybrid diesel was patented by Rudolph Diesel.

As of 1965, there were two last hopes for finding an oil province that might rival the Middle East. Those two were western Siberia and the South China Sea. As pointed out earlier, we don't yet know about the South China Sea. Western Siberia turned out to contain entirely natural gas, almost no oil. One-third of all the world's known natural gas reserves are in western Siberia.[15] The Russians do market some of that gas in Europe, but a lot of clarification of the Russian economic and political situation is required before the resource is fully utilized. As with oil from Iraq and Iran, interruption of exploitation now may preserve a portion of the resource for future generations.

Coal is the worst possible fossil fuel. Most of the fuel value comes from carbon, with the carbon dioxide added to the atmosphere. Sulfur and mercury are difficult to remove from coal; they are released to the air in ordinary burning. However, the world has at least a 300-year supply of coal. The United States and the former Soviet Union have the largest reserves of coal; the third world has comparatively little coal. That didn't seem fair, but I realized that the two largest industrial economies were originally built on coal.

China is going to be a particular problem with regard to coal burning. China has extensive coal deposits and more than a billion people. I heard a talk presented at a geology meeting that illustrated how difficult the problem can be. In one area in China there are local coal beds, but the coal contains significant amounts of arsenic. The coal is burned on raised, but open, platforms inside the homes. Slides projected at geology meetings are usually quite pleasant: interesting rocks, beautiful mountains. But the talk on arsenic-bearing coal was illustrated with horrible pictures of lesions on people's hands and feet from arsenic poisoning. The government had tried to sell the people stoves with chimneys to get the arsenic out of the house. The stoves were rejected, in part because the open fire did a better job of drying vegetables hung from the rafters. I have a suggestion that has only a slim chance of working. Often arsenic mineralization is associated with gold. If there is a decent trace of gold in that coal, I could give them the stoves and pay them for the coal ash. They avoid the arsenic poisoning and I get to keep the gold.

Our political discussions often become debates about the wrong topic. Sometimes, a subsidiary issue is a proxy for a deeper concern:

- Preserving ANWR, not because we love caribou but because we don't want the oil companies to get rich from our public lands.
- Opposing disposal plans for radioactive wastes because of our fears of nuclear power plants.

Discussions about increasing the supply of crude oil get sidetracked into debates about whether government action is needed or whether the invisible hand of economics will guide us to bigger and better oil fields. We can argue endlessly about the details without asking first whether searching for additional crude oil would be worth the effort.

When my parents retired to a farm in Oklahoma, they hired a bulldozer and built a pond. Stocked it with bass. Good fishing, good eating. Gradually, however, they began catching fewer bass, and catching enough for dinner required longer and longer hours. There are two possible reactions:

- Buy ever more expensive fishing tackle, because there might be a great lunker of a bass still hiding deep in the pond.
- Substitute fish from the grocery store and take up something other than fishing for a hobby.

The finite supply of world oil is, in my opinion, written in stone. It's not engraved on the facade of the Treasury Building. It's written in the reservoir rocks, in the source rocks, and in the cap rocks. No amount of fancy fishing tackle is going to satisfy our appetite for oil.

CHAPTER 10

Alternative Energy Sources

There are plenty of energy sources other than fossil fuels. Running out of energy in the long run is not the problem. The bind comes during the next 10 years: getting over our dependence on crude oil. This chapter begins by discussing two nonrenewable energy sources, followed by the renewable resources.

"Geothermal" energy is just what its name implies, heat recovered from within the Earth's crust. Just as with oil fields, there are a few high-grade geothermal areas, a larger number in the middle, and extensive low-grade heat sources that may not be economic in this century. Before 1960, three high-grade geothermal fields were generating electric power: Lardarello, Italy; the Geysers, north of San Francisco; and Wairakei, New Zealand (pronounced wy-RACK-ie). Each area was originally identified from hot springs at the surface, and ordinary oil-well-drilling technology was used to exploit the resource.[1] The water from these three areas was hot enough to boil, giving an impressive yield of steam. Passing the steam through turbines generated electrical power; the steam was condensed back to water on the downstream side of the turbine to increase the energy yield. The Geysers geothermal field produces about half of the electricity used by the city of San Francisco.

There are some additional high-temperature geothermal fields, but none of them seems to be as attractive as the first three. However,

A geothermal area about 80 miles north of San Francisco, called the Geysers, generates enough electricity to supply about half of the city of San Francisco. © Bob Rowan; Progressive Image/CORBIS.

stepping down in both temperature and size opens up a large number of geothermal fields. Even though the water temperature in these lesser deposits is hot enough to boil, letting 10 percent of the water convert to steam cools the rest of the water down below the boiling temperature. Also, boiling extracts a price: dissolved minerals come out of the water and crud up the equipment. The older high-temperature deposits put up with the problem; Lardarello even sold boron chemicals as a by-product.[2] For a leaner deposit, the cost of the labor and downtime for cleaning the pipes is unacceptable.

A clever process for recovering geothermal energy without the consequences of boiling became commercial during the 1980s. Hot water from geothermal wells was used to boil an organic liquid, but the water itself was returned back to the ground without being allowed to boil. The organic vapor goes through turbines to generate electricity, and the vapor is condensed back to liquid and circulated to con-

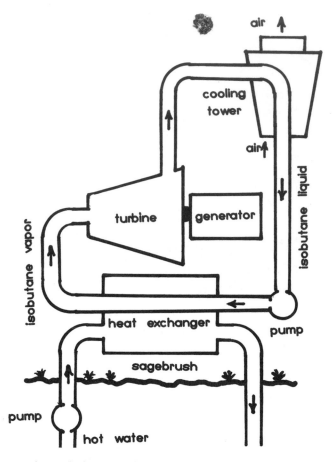

A binary geothermal plant uses hot water from the ground to boil an organic liquid. The water is returned to the ground, and the organic vapor goes through a turbine to generate electricity.

tact more hot water. These are called "binary" geothermal plants because both water and a low-boiling liquid are used.

The beauty of a binary geothermal plant is zero emissions to the atmosphere and returning the water to the ground. You don't need an Environmental Protection Agency permit to pump arsenic, mercury, and antimony back into the ground if you just took the water out of an adjacent hole. The binary process opens a range of geothermal opportunities that would not otherwise be economic. Since the late

1980s, I have been taking 20 first-year students during fall break to study geology around Mammoth Lakes, California. One stop on the trip is at an elegantly designed and profitable binary geothermal plant. Hot water from wells drilled into a welded tuff (with connected pores) is used to boil isobutane, which boils at 11°F. One of my students asked the plant engineer what would happen if one of the geothermal wells got out of control. He said that he would call a company named Halliburton. Right! Call Duncan, Oklahoma. The engineer also pointed out that the plant, over its lifetime, would replace 3 million barrels of imported oil. After he left, I had to tell the students that the United States was importing 6 million barrels of oil a *day*. That plant solves the U.S. energy problem for one half of one day.

Drilling for geothermal energy utilizes the same equipment and skills that were developed for the oil industry. It comes as no surprise that oil companies looked at geothermal resources as an extension of their existing activities. There are rumors that at least one major oil company is holding leases on U.S. geothermal areas as a way of extending its business into an era of energy shortage.

Everywhere, the temperature in the Earth's crust increases with depth; that observation underlies the oil-window paradigm. In a sense, there is geothermal energy everywhere. The U.S. Department of Energy sponsored one test to get energy out of hot dry rock.[3] At the edge of a major volcanic area, west of Los Alamos, two vertical holes were drilled, and oil field hydrofrac equipment was used to open a fracture between the two holes. Cold water was pumped down one hole, heated by the rock, and hot water came up the other hole. When I was taken on a group tour to see this operation, someone in the audience asked whether the hot water coming up carried more energy than the pumps consumed. The tour guide didn't know which was larger, but he gave the pump pressure in pounds per square inch, the flow rate in gallons per minute, and the temperatures in degrees. A geophysicist, whose work I greatly admired, turned around and started mumbling; he was trying to work the problem in his head. Game time! Could I beat him to the answer? I had recently done some similar calculations for another purpose. We finished at almost the same time.

(I like to think that I was one second ahead.) The hot water would not produce enough energy to run the pumps.

Drilling for geothermal energy has its unpleasant surprises. The federal government and the state of California jointly sponsored a geothermal test right in the center of the Mammoth Lakes–Long Valley volcanic area. Nearby volcanoes are only 400 years old. A drilling rig was brought in. Whenever I took my students to see it, I described the rig as "medium large." In truth, that particular rig held the American depth record: drilled beyond 30,000 feet in the Anadarko Basin of Oklahoma. The Mammoth Lakes "geothermal" hole got down to 12,000 feet, and the drillers were not even in volcanic rocks. They were drilling cold dry rock typical of the adjacent Sierra Nevada. A year ago the rig was cut up and sold for scrap iron.

Today, nuclear energy is about as unfashionable as the hoop skirt. The incidents at Three Mile Island and Chernobyl converted a widespread uneasy feeling into an almost universal fear of nuclear power plants. On other side of the nuclear argument: no carbon dioxide emission to the atmosphere and a 100-year supply of uranium. It's going to be a gut-wrenching debate. Are painfully high electric bills the cure for nuclear phobia?

No press agent would have advertised nuclear power by blowing away Hiroshima and Nagasaki. It could have happened another way. In 1972, the French found that uranium from the Oklo Mine in Gabon was missing most of its fissionable uranium 235. After some investigation, they established that a natural nuclear fission reaction had occurred.[4] In effect, a billion years ago in nature there were six (count them, six) water-moderated, enriched-uranium nuclear reactors. All natural ingredients. The "enriched" uranium existed because a billion years ago less of the uranium 235 had decayed. Suppose the natural reactors at Oklo had been discovered in 1935 instead of 1972. Wow! A wonderful gift from nature. Best thing since Prometheus discovered fire. The nuclear age would have opened with a whimper instead of a bang.

Earlier, I mentioned the grip that the internal combustion engine has on the automobile and the rotary drilling rig has on the oil business. Similarly, the enriched-uranium, water-moderated nuclear reactor dominates the market for nuclear power plants. Actually, the standard commercial nuclear power reactors are derived from the U.S. Navy nuclear submarine reactors. However, there are about a dozen fundamentally different designs for nuclear reactors.[5] I once heard Eugene Wigner lecture on reactor designs. During World War II, Wigner designed the reactors at Hanford; later he won the Nobel Prize for bringing group theory into physics. Back when new power reactors were being built, the Canadians built and operated several reactors that used natural uranium and "heavy water" (deuterium) as the moderator that slowed down the neutrons. The Canadian deuterium reactors went by the overly clever name of CANDU. The CANDU reactors had an outstanding track record for reliability. I suspect that if we sat down today to reinvent the nuclear power industry, we might not choose the standard American design.

Reactors have two options for their used uranium. The spent fuel elements can be disposed of in their entirety as radioactive waste. Or the used uranium can be "reprocessed" to recover unburned uranium and to recover plutonium. Plutonium is evil stuff. Besides being exceedingly toxic, plutonium can be used as the core of a nuclear bomb. Ted Taylor, who was the leading American designer of nuclear weapons, left Los Alamos to campaign against nuclear proliferation because he felt that it was entirely too easy to build an effective nuclear bomb out of stolen plutonium. There is a continuing debate about whether you need a large staff of whizzbang scientists to build a nuclear weapon or whether you could build one in your garage. No question that Ted Taylor could build one in his garage.

Despite the scary aspect, some spent fuel from commercial power reactors has been reprocessed and the plutonium recycled for additional reactor fuel. A dilemma over reprocessing arose for the Australians around 1980. Huge uranium deposits had just been discovered in northern Australia; the Jabilinka deposit by itself contained more uranium that the United States had mined since 1940.[6] The Australians preferred not to encourage worldwide growth of nuclear power plants. But as the specter of reprocessing and commercial plutonium shipments arose, the Australians switched their position. If Australian uranium is cheap enough, nobody will want to reprocess spent fuel.

Very likely the safety problems and the radioactive waste problems could be handled adequately if our society were willing to make the appropriate investments. Maybe the hoop skirt will come back. The emotional problem remains. There is one possible psychological boost. The United States has been converting surplus weapons-grade uranium and plutonium, some of it purchased from the Russians, into reactor fuel. Would getting rid of nuclear weapons be an incentive for accepting nuclear power plants? "They shall beat their swords into plowshares and their spears into pruning hooks. They shall burn their warheads to warm their toilet seats." OK, OK, so Ecclesiastes was a better writer than I am; it's the thought that counts.

It is difficult to store electrical power. There are a few facilities that use electricity to pump water uphill to a reservoir at times of low demand and generate power when needed by running the water back down. Only a few sites are suitable for pumped storage; most electric power is generated at the time it is needed. Typically, nuclear power plants carry the base load, the minimum power demand that exists all the time. At the intermediate level, hydroelectric and fossil-fuel steam-generating plants carry the normal increase. For the few peak hours, a hot summer afternoon with all the air-conditioners running, the preferred power source requires a low capital cost but can tolerate a high fuel cost. A gas turbine, similar to a jet aircraft engine, powered by oil or natural gas is a typical peak-load power generator. Some of the renewable electric sources, like solar and wind power, have their own built-in schedules; the difficulty of storing power is a serious problem.

Hydroelectric power has a number of attractions. It is renewable, you can turn it on when it is needed, and it does not pollute the atmosphere. Enthusiasm for hydropower was enormous during the first half of the twentieth century. Wild rivers, salmon runs, and the glens of Glen Canyon were sacrificed in exchange for cheap electric power. Of course, mountainous areas have the greatest hydroelectric opportunities. Hydroelectric power is the leading export of Kyrgyzstan; most of the country consists of the Tien Shan mountain range. Locally generated hydropower supplies electricity in remote Himalayan valleys. Aluminum ore is hauled halfway around the world to get to cheap hydroelectric power at Kitimat, British Columbia.[7] Right now, there is a growing enthusiasm for tearing down dams, instead of building more. In mountainous areas with high rainfall, new hydroelectric facilities might be acceptable, but the total of all the remaining hydroelectric opportunities will make only a modest dent in the need for electricity.

Solar and wind power participate in what I call the energy-material paradox. If materials were cheap, I could build large energy collectors. If energy were cheap, I could produce large amounts of raw materials. If neither materials nor energy is cheap, I have a problem. At the moment, solar and wind power are developing in specialized areas. Neither is an immediate, large-scale solution to the energy problem.

The power per square foot in the sunshine is essentially identical to the power per square foot in the wind. At first, I thought that this was just an accident, but solar energy may crank up the wind velocity until the average energy density in the wind equals the average solar energy density. Wind and sun don't appear together at the same place or the same time. As an example: wind often is stronger at a gap in the mountains. At the southern end of the Sierra Nevada, there is a huge array of windmills. They don't look like Dutch windmills, or like my grandmother's daisy wheel; they are slender, three-bladed propellers on tall stands.

The low energy density in solar and wind power requires large energy collectors. A normal-size nuclear or fossil-fuel power plant generates 1,000 megawatts. At typical efficiencies around 10 percent, a solar or wind collector has to occupy five square miles to deliver 1,000 megawatts. I can direct you to any of several Nevada basins where you can get the five square miles; your problem is the capital cost of paving five square miles with solar collectors.

During the 1980 energy crisis, a variety of different ways were considered for converting solar energy to electric power. The two major approaches were:

The engineering design for generating electricity from windmills is not as simple as it might sound. After the engineering problems were solved, large arrays of windmills could be built by replicating a single design. © Bob Rowan; Progressive Image/CORBIS.

1 Trap solar heat and use the heat to generate electricity. For instance, boil a liquid with solar heat and run the vapor through a turbine. High efficiency requires a large temperature difference between the solar-heated source and the air or water where the waste heat is dumped.

2 Direct generation of electricity, typically by using semiconductor solar cells. In 1980, the efficiency of solar cells was about 6 percent; today's cells are about 13 percent efficient.

The materials for solar and wind collectors are not scarce. Silicon (for solar cells) and aluminum (for windmills) are the second and third most abundant elements in the Earth's crust. However, producing either silicon or aluminum from their ores requires lots of energy. It's the energy-material dilemma that was mentioned earlier.

During the 1980 oil crisis, exotic energy schemes popped up like dandelions in my lawn. Examples are (1) solar cells in Earth orbit, sending down solar power by microwave, and (2) exploiting the temperature difference between warm surface seawater and cold deep seawater. None of the schemes are to be ignored, but the inventor usually wants to see an immediate full-scale implementation of the idea. It takes real judgment to sort the sheep from the goats. It is all too easy to say, "It won't work." Rather than make decisions in a closed committee, we need an open competition to propose detailed implementations of each concept, followed by another competition to look for serious flaws in the designs.

CHAPTER 11

A New Outlook

During the 2000 presidential campaign, Democrats and Republicans debated about how to use the new surplus in the federal budget: pay off the national debt, fix Social Security, improve Medicare, or reduce taxes. There is another option: gift wrap the entire surplus and present it to the Saudi royal family. We could go happily on, pretending that either (1) a permanent decline in world oil production won't happen or (2) it doesn't matter. Ask anyone who remembers the 1980 crisis: it happens and it matters. In 1980 it was a problem in distribution; the oil was there, but it wasn't getting to the corner gas station. In 2008, the oil won't be there. The psychological realization that the change is permanent may be as devastating as the shortage itself.

Aren't we seeing the dawning of the age of the Internet? Isn't the computer the second Industrial Revolution? The computer doesn't use much energy. For 10 cents' worth of electricity, my desktop computer will do a trillion multiplications. A million million operations for a dime. Fiber optic communications use so little energy that I can shoot e-mail messages halfway around the world and my Internet service provider charges only a flat fee per month. Is it ridiculously old-fashioned to be worrying about oil when the computer brings us an enormous gain in productivity? Legend has it that a bright young economist said in his lecture that we shouldn't worry about things like agriculture and mining, as they make up only 3 percent of the U.S.

gross national product. An older person in the back of the room muttered, "What does that genius think we are going to eat?" If we take the experiences of 1967, 1973, and 1980 as guides, what lessons have we learned?

First, beware of any salesman peddling just one brand of snake oil. There will be numerous voices claiming to have the new, new thing to solve the energy problem. They are not necessarily con artists. Some of them convince themselves first, then they try to con the rest of us. They are their own first victims. We should make good use of each innovation where it fits best. Use geothermal energy where it is most effective; don't try to find a geothermal solution for the entire U.S. energy needs.

Second, beware of the salesman peddling an enormous variety of snake oils. His message is, "There are *so* many possibilities, some of them are bound to come through in time to save us." Usually a long list of innovations, including gas hydrates, subsalt seismic reflections, coal bed methane, and deep-water drilling, gives the impression that doomsday won't arrive in our lifetime. We'll muddle through. Unfortunately, the items in that list were already identified 20 years ago. It may be a painful muddle.

There are some possibilities for doing a better job than we did in 1980. Rather than have the crisis sneak up on us, we can see it coming and initiate some of the long lead-time projects in advance. "Forewarned is forearmed."

In earlier oil shortages, there was a huge fixation on improving automobile efficiencies in miles per gallon. Unfortunately, some of those improved miles per gallon came at a cost of higher energy consumption in the refinery.[1] We need to look at the whole system, including the refinery and the car manufacturer, when we "improve" the system.

There is another kind of "efficiency" that is misleading. When I switched my house from oil to natural gas, I paid extra for a high-efficiency furnace. In addition to extracting heat from burning the gas, the furnace condenses the water vapor from the combustion to supply additional heat to the house. I can stick my hand right in the

It may seem easy to increase the fuel efficiencies of automobiles by changing the specifications for gasoline and diesel fuel. If the required changes increase the energy used inside the refinery, the change may be self-defeating. © Bettmann/CORBIS.

exhaust pipe without getting burned. The efficiency is listed at 94 percent. Sounds fabulous—hard to improve on 94 percent.

Here's the problem: just downstream from the gas-air flame, the temperature is over 1,000 degrees. It's high-grade energy. My "efficient" furnace ignores the high quality and dilutes everything down to put hot air in the ducts. The alternative is cogeneration: generate electricity using the high-temperature gases, then use the lower temperatures to heat the house. The smallest cogeneration units operate on the scale of an apartment block.

We need to approach energy use in the same spirit in which a butcher approaches a side of beef. He could convert all the meaty part to hamburger, with high efficiency. Butchers (and economists) know to slice off some high-priced steak and a roast or two, then grind up the rest for hamburger. Electricity is steak; space heating is hamburger.

Analyzing thermal efficiencies is a science, largely developed in the nineteenth century, known as "thermodynamics." "Thermo" as the students call it, is a fundamental field.[2] In Einstein's opinion, of all our physical laws, classical thermodynamics could be the part that will never be overthrown. About 20 years ago, the *New York Times* estimated that there were fewer than 100,000 people in the entire world who knew something that they could use about thermodynamics. If you are among the 100,000, the 94 percent for my furnace is a "first-law efficiency." Cogeneration tries to optimize the "second-law efficiency."

The simplest reason for recycling is reducing the load on the mining industry. For copper, it is straightforward; any copper we recycle is copper we don't have to mine. Aluminum metal costs about $200 per ton, but that is $3 for the aluminum ore and $197 for electricity. The motivation for recycling aluminum is energy conservation. Most of the glass bottles we recycle get remelted into new glass bottles. Remelting glass is energy-intensive. When I was young, milk bottles and the signature Coca-Cola bottles were washed and refilled. We might reinvent reuse, rather than remelting.

I have mentioned more than once our need to get over our nuclear phobia. Hubbert's 1956 paper,[3] with his famous prediction, is entitled "Nuclear Energy and the Fossil Fuels." We certainly need better engineering and better operators. A nuclear plant operator ought to receive the level of training and the salary of an airline pilot.

Awareness is important. Of course, the economic squeeze will get everyone's attention. The experience that raised my awareness was a bicycle frame hitched to an electric generator wired to a light bulb. You could switch on a 50-watt bulb, pedal the bicycle, and keep the light lit. Change to a 100-watt bulb and it took a sustained serious effort to keep the bulb glowing. I couldn't light up a 200-watt bulb. It put a real scale on energy conservation.

In May 2000, the president of the American Association of Petroleum Geologists wrote an optimistic evaluation of the future and concluded with this statement:

> "What do I tell my grandchildren, ages 8, 3, and 2 . . . about the future of a geoscience career?
>
> "I will tell them that if they (1) are looking for an exciting career where you can use your brains to 'play' with constantly improving technology and (2) enjoy the satisfaction of making a major contribution to the society in which they live—they can't beat a professional career as a geologist.
>
> "So to Justin, Jakob, and Mollie: 'There will be a career in geoscience for you if you work hard enough—and it will be a wonderful life for each of you.'"[4]

I have some advice for my granddaughter, age two:

> Learn something that you can use about thermodynamics. By the time you reach retirement age, Emma, world production of oil (the kind that's fun to drill for) will be down to a fifth of its present size.
>
> Get into renewable energy. Look at a cornstalk the way a Chicago meatpacker used to look at a hog: sell everything but the squeal. If you need some oil-based lubricants in your bio-energy-pharmaceutical plant, ask whether Justin, Jakob, and Mollie are interested in scratching for the last few strat traps in Iraq.
>
> Love,
> Grandpapa

NOTES

──────────────── *Chapter 1* ────────────────

1. Hubbert, M. K. (1956), "Nuclear Energy and the Fossil Fuels," American
    Petroleum Institute Drilling and Production Practice, Proceedings of
    Spring Meeting, San Antonio, 1956, pp. 7–25; also Shell Development
    Company Publication 95, June 1956.
2. Hatfield, C. B. (1997), "Oil Back on the Global Agenda," Nature **387**:121;
    Kerr, R. A. (1998), "The Next Oil Crisis Looms Large—And Perhaps
    Close," Science **281**:1128–31; Campbell, C. A., and J. H. Laherrere
    (1998), "The End of Cheap Oil," Scientific American, March:78–83.
3. Several search engines, several databases, and several librarians did not
    turn up this title. Please e-mail deffeyes@princeton.edu if you locate the
    original source.
4. Campbell, C. J. (1997), The Coming Oil Crisis, Multi-Science Publishing
    Company and Petroconsultants. His world scenario is on p. 201.
5. Hubbert, M. K. (1981), "The World's Evolving Energy System," American
    Journal of Physics 49:1007–29.
6. Yergin, Daniel (1991), The Prize, New York: Simon & Schuster. The
    founding of OPEC is described beginning on p. 522.
7. In this book, the reserves and production data are those reported by the
    Oil and Gas Journal in its last issue of each year. Campbell's evaluation
    of OPEC reserve increases is on p. 73 in his Coming Oil Crisis.
8. Colin Campbell and Sam Carmalt have worked for, or with, Petro-
    consultants. In 1998, Petroconsultants was merged into the IHS Energy
    Group, www.ihsenergy.com.
9. One of the best critical rejections of Hubbert's approach is Adelman, M.
    A., and M. C. Lynch (1997), "Fixed View of Resource Limits Creates Un-

due Pessimism," Oil and Gas Journal, April 7:56–60. Other comments are in Oil and Gas Journal, February 23, 1998:77, and November 2, 1998:94.

10. As explained in the next chapter, David White of the U.S. Geological Survey was a pioneer in interpreting the origins of oil. In 1929, White published "Description of Fossil Plants Found in Some 'Mother Rocks' of Petroleum from Northern Alaska," American Association of Petroleum Geologists Bulletin **13**:841–48.

11. McPhee, John (1998), Annals of the Former World, New York: Farrar, Straus, and Giroux. This is a revised printing of four earlier books.

12. McCaslin, J. C. (1984), "Well Completions Boost Impact in New York Area," Oil and Gas Journal, March 5:121–22.

———————————— *Chapter 2* ————————————

1. Kornfeld, J. A. (1962), "Tidelands Drilling Begins in Washington-Oregon Area," World Oil **155**, no. 4:89–90.

2. Philippi, G. T. (1957), "Identification of Oil Source Beds by Chemical Means," Report of the 20th Session of the International Geological Congress, Mexico City, Section 3, pp. 25–38. A later version (1965, "On the Depth, Time, and Mechanism of Petroleum Generation") appeared in Geochimica et Cosmochimica Acta **29**:1021–49.

3. Krauss, K. G., et al. (1997), "Hydrous Pyrolysis of New Albany and Phosphoria Shales," Organic Geochemistry **27**:477–96.

4. Ayres, M. G., et al. (1982), "Hydrocarbon Habitat in Main Producing Areas, Saudi Arabia," American Association of Petroleum Geologists Bulletin **66**:1–9.

5. Davis, H. R., et al. (1989), "Depositional Mechanisms and Organic Matter in Mowry Shale," American Association of Petroleum Geologists Bulletin **73**:1103–16.

6. Pines, H. (1981), "The Chemistry of Catalytic Hydrocarbon Conversions," New York: Academic Press.

7. Kandel, E. R., et al. (2000), Principles of Neural Science, 4th ed., New York: McGraw-Hill, p. 82.

8. Anders, D. E., and W. E. Robinson (1973), "Geochemical Aspects of the Saturated Hydrocarbon Constituents of Green River Oil Shale," U.S. Bureau of Mines, Report of Investigations 7737.

9. McPhee, J. A. (1998), Annals of the Former World, New York: Farrar, Straus, and Giroux, pp. 174–78.

10. World Oil Staff (2001), "Drilling and Producing Depth Records," World Oil, February:71–73.

11. White, G. T. (1968), Scientists in Conflict, San Marino, Calif.: Huntington Library.

12. Brennan, P. (1990), "Greater Burgan Field," in American Association of

Petroleum Geologists, Tulsa, Treatise on Petroleum Geology, vol. A-106, pp. 103–28.

13. The quote is the closing line of "The Woodpile."

14. Judson, S., et al. (1976), Physical Geology, Englewood Cliffs, N.J.: Prentice-Hall. The silled-basin diagram is on p. 435.

15. Demaison, G. J., and G. T. Moore (1980), "Anoxic Environments and Oil Source Bed Genesis," Organic Geochemistry 2:9–31.

16. White, D. (1915), "Geology: Some Relations in Origin between Coal and Petroleum," Journal of Washington Academy of Sciences 5:189–212.

17. Whelan, J. K., and C. Thompson-Rizer (1993), "Chemical Methods for Assessing Kerogen and Protokerogen Types and Maturity," in Organic Geochemistry, ed. M. H. Engel and S. A. Mako, New York: Plenum Press, pp. 289–353.

18. McPhee, Annals of the Former World. The discussion of the origin of oil begins on p. 178.

19. Rejebian, V. A. (1987), "Conodont Color and Textural Alteration," Bulletin of the Geological Society of America 99:471–79.

20. Love, J. D., et al. (1961), "Relation of Latest Cretaceous and Tertiary Deposition and Deformation to Oil and Gas Occurrences in Wyoming," American Association of Petroleum Geologists Bulletin 45:415.

21. Morgan, W. J. (1980), "Hotspot Tracks in North America," Eos, Transactions, American Geophysical Union 61:380.

22. Epstein, A. G., et al. (1977), "Conodont Color Alteration," U.S. Geological Survey Professional Paper 995.

23. Dana, J. D. (1997), Dana's New Mineralogy, rev. 8th ed., New York: John Wiley & Sons. Faujasite localities are on p. 1660.

24. Mair, B. J. (1964), "Hydrocarbons Isolated from Petroleum," Oil and Gas Journal September 14:130–34.

25. Deffeyes, K. S. (1982), "Geological Estimates of Methane Availability," in Methane, Fuel for the Future, ed. P. McGeer and E. Durbin, New York: Plenum Press, pp. 19–29.

26. Pines, "Chemistry of Catalytic Hydrocarbon Conversions."

——————————— *Chapter 3* ———————————

1. Owen, E. W. (1975), "The Trek of the Oil Finders," American Association of Petroleum Geologists Memoir 6.

2. Hubbert, M. K. (1953), "Entrapment of Petroleum under Hydrodynamic Conditions," American Association of Petroleum Geologists Bulletin 37:1954–2026.

3. Minor, H. E., and M. A. Hanna (1941), "East Texas Oil Field," in Stratigraphic Type Oil Fields, ed. A. I. Levorsen, Tulsa: American Association of Petroleum Geologists, pp. 600–640.

4. Hutton, J. (1899), Theory of the Earth, vol. 3, reprinted by Scholars' Facsimiles and Reprints, Delmar, N.Y., p. 235.

5. Smith, D. A. (1980), "Sealing and Nonsealing Faults in Louisiana Gulf Coast Salt Basin," American Association of Petroleum Geologists Bulletin **64**:145–72.

6. American Commission on Stratigraphic Nomenclature (1961), "Code of Stratigraphic Nomenclature," American Association of Petroleum Geologists Bulletin **45**:645–60.

7. New York Times, February 24, 1988, sec. A, p. 1, col. 6.

8. Dunham, R. J. (1970), "Stratigraphic Reefs versus Ecologic Reefs," American Association of Petroleum Geologists Bulletin **54**:1931–32.

9. Andrichuk, J. M. (1958), "Stratigraphic and Facies Analysis of Upper Devonian Reefs in Leduc, Stettler and Redwater Areas," American Association of Petroleum Geologists Bulletin **42**:1–93. Replacement of the original limestone by dolomite has obscured most of the original textures in the Leduc reefs; evaluating the original reef ecology is not possible.

10. Bass, N. W. (1934), "Origin of the Bartlesville Shoestring Sands of Greenwood and Butler County, Kansas," American Association of Petroleum Geologists Bulletin **18**:1313–45.

11. Shirley, K. (1984), "Point Bars Stir Los Animas Activity," American Association of Petroleum Geologists Explorer **5**:42–43.

12. Murray, D. K., and L. C. Bortz (1967), "Eagle Springs Oil Field," American Association of Petroleum Geologists Bulletin **51**:2133–45.

13. Henry, C. D., et al. (1997), "Brief Duration of Hydrothermal Activity at Round Mountain, Nevada," Economic Geology **92**:807–26.

14. Hubbert, M. K. (1963), "The Physical Basis of Darcy's Law," Journal of Petroleum Technology **15**:849.

15. Davies, D. K. (1966), "Sedimentary Structures and Subfacies of a Mississippi River Point Bar," Journal of Geology **74**:234–39.

16. McKee, E. D. (1979), "A Study of Global Sand Seas," U.S. Geological Survey Professional Paper 1052.

17. McKerrow, W. S., and F. B. Atkins (1989), Isle of Arran, 2d ed., Geologists' Association, p. 61.

18. Blatt, H., et al. (1980), Origin of Sedimentary Rocks, 2d ed., Englewood Cliffs, N.J.: Prentice-Hall, pp. 176–82.

19. Hsu, K. J. (1977), "Studies of Ventura Field, California," pts. 1 and 2, American Association of Petroleum Geologists Bulletin **61**:137–91.

20. Arabian American Oil Company Staff (1959), "Ghawar Oil Field, Saudi Arabia," American Association of Petroleum Geologists Bulletin **43**:434–54.

21. Krumbein, W. C., and L. L. Sloss (1963), Stratigraphy and Sedimentation, 2d ed., San Francisco: W. H. Freeman, p. 176.

22. Van Tuyl, F. M. (1916), "The Origin of Dolomite," Iowa Geological Survey Bulletin **25**:241–422, and (1916), "New Points on the Origin of Dolomite," American Journal of Science **42**:249–60.

23. Saller, A. H., and N. Henderson (1998), "Distribution of Porosity and Permeability in Platform Dolomites: Insight from the Permian of West Texas," American Association of Petroleum Geologists Bulletin **82**:1528–50.

24. Deffeyes, K. S., et al. (1964), "Dolomitization: Observations on the Island of Bonaire," Science **143**:678–79.

25. Sneider, R. M., et al. (1997), "Comparison of Seal Capacity Determinations," pp. 1–12 in American Association of Petroleum Geologists Memoir 67.

26. Wasserburg, G. J., and E. Mazor (1965), "Spontaneous Fission Xenon in Natural Gases," pp. 386–98 in American Association of Petroleum Geologists Memoir 4.

27. Chaturvedi, L., et al. (1996), "Issues in Predicting the Long-Term Integrity of the WIPP Site," Eos, Transactions, American Geophysical Union **77**:F19–F20.

──────────────── *Chapter 4* ────────────────

1. Adelman, M. A., and M. C. Lynch (1997), "Fixed View of Resource Limits Creates Undue Pessimism," Oil and Gas Journal, April 17.

2. New York State maintains public log libraries, with coin-operated photocopy machines. Most other areas rely on commercial log reproduction services.

3. Allaud, L. A., and M. H. Martin (1977), Schlumberger, the History of a Technique, New York: John Wiley.

4. Serra, O. (1985), Sedimentary Environments from Wireline Logs, Houston: Schlumberger, pp. 103 and 134.

5. Beaton, K. (1957), Enterprise in Oil: A History of Shell in the United States, New York: Appleton Century Crofts, p. 646.

6. Archie's first law states that the ratio of the electrical resistivity (ohms/meter) of a sedimentary rock whose pores are filled with salt water, R_r, to the resistivity of the water, R_w, depends on 1 divided by the square of the porosity, Φ. The porosity is expressed as a fraction; for 10 percent porosity, $\Phi = 0.1$:

$$\frac{R_r}{R_w} = \frac{1}{\Phi^2}$$

Archie's second law states that the ratio of resistivity of a rock whose water is partially replaced with oil or gas, R_t, to the resistivity of the same rock fully saturated with water, R_w, depends on 1 divided by the fraction of the pore space occupied by water, S_w:

$$\frac{R_t}{R_r} = \frac{1}{S_w^2}$$

7. Archie's first and second laws can be combined by eliminating R_r between them, solving for S_w, and recognizing that the fraction of the pore space occupied by oil or gas, S_o is 1 minus S_w:

$$S_o = 1 - \frac{1}{\Phi} \sqrt{\frac{R_w}{R_t}}$$

8. A current summary of Schlumberger services is at www.slb.com.
9. Dobrin, M. B., and C. H. Savit (1988), Introduction to Geophysical Prospecting, 4th ed., New York: McGraw-Hill, p. 573.
10. Sweet, G. E. (1978), The History of Geophysical Prospecting, Los Angeles: Science Press, p. 81.
11. Petty, O. S. (1976), Seismic Reflections, Houston: Geosource, p. 21.
12. Texas Instruments corporate history is available at www.ti.com/corp/docs/company/history. The company conducted reflection seismic surveys from 1930 to 1988. Jack Kilby invented the integrated circuit in 1958.
13. Geyer, R. L. (1989), Vibroseis, Tulsa: Society of Exploration Geophysicists Geophysics Reprint Series No. 11. Pulse compression in radar was invented earlier by Robert Dicke, U.S. Patent 2,624,876 (1953).
14. Zoeppritz, K. (1919), "Uber Erdbebenwellen VIIb," Göttinger Nachrichten, pp. 66–84.
15. A biography of Harry Mayne, who invented CDP stacking for Petty Geophysical, is available in a "virtual museum" maintained by the Society of Exploration Geophysics: www.seg.org/museum/.
16. Forrest, M. (2000), "Bright Investments Paid Off," AAPG Explorer, July:18–21.
17. Caughlin W. G., et al. (1976), "The Detection and Development of Silurian Reefs in Northern Michigan," Geophysics **41**:646–58.
18. "Improved Satellite Transmission Speeds Challenge Conventional Wisdom on Processing Marine Seismic Data" (2000), First Break, May: 193–96. A few years ago a seismic contractor, Western Geophysical, said that its third-largest computer, after Houston and London, was on one of its ships. Now it is sending the raw data by satellite to land-based computer centers.
19. The Center for Wave Propagation at the Colorado School of Mines has been adapting its reflection seismic software for parallel computer clusters: www.cwp.mines.edu.
20. Claerbout, J. F. (1976), Fundamentals of Geophysical Data Processing, Palo Alto, Calif.: Blackwell, p. 12.
21. Claerbout, J. F. (1985), Imaging the Earth's Interior, Palo Alto, Calif.: Blackwell, p. 385.

22. Suppe, J. (1985), Principles of Structural Geology, Englewood Cliffs N.J.: Prentice-Hall, p. 57.

23. Suppe, J., and D. A. Medwedeff (1990), "Geometry and Kinematics of Fault-Propagation Folding," Eclogae Geol. Helv. **83**:409–54.

———————————— *Chapter 5* ————————————

1. Brantly, J. E. (1971), History of Oil Well Drilling, Houston: Gulf Publishing.

2. Eby, J. B., and M. T. Halbouty (1937), "Spindletop Oil Field, Jefferson County, Texas," American Association of Petroleum Geologists Bulletin **21**:475–90. The Spindletop museum has a web site at http://hal.lamar .edu/~psce/gladys.html.

3. Cannon, G. E., and R. C. Craze (1938), "Excessive Pressures and Pressure Variations with Depth in the Gulf Coast Region of Texas and Louisiana," Transactions of the American Institute of Mining and Metallurgical Engineers **127**:31–38.

4. Hottman, C. E. (1965), "Estimation of Formation Pressures from Log-Derived Shale Properties," Journal of Petroleum Technology **17**:717–22.

5. Hubbert, M. K., and W. W. Rubey (1959), "Mechanics of Fluid-Filled Porous Solids and Its Application to Overthrust Faulting," Geological Society of America Bulletin **70**:115–66 and 167–205.

6. Verbeek, E. R. (1977), "Surface Faults in the Gulf Coast Plain between Victoria and Beaumont, Texas," Tectonophysics **52**:373–75.

7. Osborne, M. J., and R. E. Swarbrick (1997), "Mechanisms for Generating Overpressure in Sedimentary Basins," American Association of Petroleum Geologists Bulletin **81**:1023–41.

8. Deffeyes, K. S. (2001), "Overpressures from Mineral Dehydration," European Union of Geosciences XI, Strasbourg, p. 247. The key equation is

$$\sigma = \frac{-\Delta H \log(T/T_o)}{V_w}$$

σ = effective stress

V_w = molar volume of water

ΔH = molar enthalpy change

T = absolute temperature

T_o = 1 atm. dehydration temp.

9. Jowett, E. C., et al. (1993), "Predicting the Depths of Gypsum Dehydration in Evaporitic Sedimentary Basins," American Association of Petroleum Geologists Bulletin **77**:402–13.

10. Buxtorf, A. (1916), "Prognosen und Befund Bein Hauensteinbasis-und Grenchenbergtunnel und die Bedeutune der letztern fuirdie Geologie des Juragebirges," Verh. Naturforsch. Ges. Basel **27**:185–254.

11. Eslinger, E., and D. Pevear (1988), "Clay Minerals for Petroleum Geolo-

gists and Engineers," Short Course Notes No. 22, Society for Economic Paleontologists and Mineralogists, Tulsa (paginated by chapters).

12. Law, B. E., et al., eds. (1988), "Abnormal Pressures in Hydrocarbon Environments," American Association of Petroleum Geologists Memoir 70, 264 pp.

13. Suppe, J., and J. H. Wittke (1977), "Abnormal Pore-Fluid Pressures in Relation to Stratigraphy and Structure in the Active Fold-and-Thrust Belt of Northwestern Taiwan," Petroleum Geology of Taiwan **14**:11–24.

14. Brown, K. M. (1994), "Fluids in Deforming Sediments," in The Geological Deformation of Sediments, ed. A. Maltman, London: Chapman & Hall, pp. 205–37.

15. Temperatures high enough to initiate clay dehydration and overpressuring are approximately the temperature at the bottom of the oil window.

16. Higham, C. (1993), Howard Hughes, New York: Putnam's Sons, pp. 17–30. More than I wanted to know about Howard Hughes.

17. Pitzer, K. S., and L. Brewer (1961), Thermodynamics, New York: McGraw-Hill, p. 100.

18. Steinhart, C. E. (1972), Blowout, a Case Study of the Santa Barbara Oil Spill, North Scituate, Mass.: Duxbury.

19. Halliburton has a corporate history posted at www.halliburton.com/whoweare/about_background.asp.

20. Britt, L. K. (2000), "Fracturing and Stimulation Overview," Journal of Petroleum Technology, March:24.

21. Swenson, D. V., and L. M. Taylor (1982), "Analysis of Gas Fracture Experiments Including Dynamic Crack Formation," Report SAND82-0633. Available from National Technical Information Service, Springfield, Va.

22. U.S. Patent 6,173,776, issued January 16, 2001.

23. Dake, L. P. (1978), Fundamentals of Reservoir Engineering, Amsterdam: Elsevier, p. 80.

24. The 1961 version of the story referred to gas recycling through the Second Wall Creek Sandstone in the Salt Creek oil field, Wyoming.

25. Peaceman, D. W. (1977), Fundamentals of Numerical Reservoir Simulation, Amsterdam: Elsevier, p. 46.

26. Donaldson, E. C. (1985 and 1989), Enhanced Oil Recovery, Amsterdam: Elsevier, in 2 vols.

27. Lewis, J. P., et al. (2000), "Improving PDC Performance, Prudhoe Bay, Alaska," Journal of Petroleum Technology, December:34–35.

28. Van Venrooy, J., et al. (2000), "Underbalanced Drilling with Coiled Tubing in Oman," Journal of Petroleum Technology, February:30–31.

29. Taylor, R. W., and R. Russell (1998), "Multilateral Technologies Increase Operational Efficiencies in Middle East," Oil and Gas Journal, March 16:76–80.

30. Rasmus, J., et al. (2000), "Logging-While-Drilling Azimuthal Measurements Optimize Horizontal Laterals," Journal of Petroleum Technology, September:56–60.

──────────────── *Chapter 6* ────────────────

1. Menard, H. W., and G. Sharman (1975), "Scientific Use of Random Drilling Models," Science **190**:337–43.
2. Yearly compilations of exploration and production results are compiled by the International Oil Scouts Association and published by the Mason Map Service, Austin, Tex. What we cheerfully call "oil scouts" would be classed as industrial spies in any other industry.
3. Bloomfield, P., et al. (1979), "Volume and Area of Oilfields and Their Impact on Order of Discovery," report to the Department of Energy for contract EI-78-S-01-6540 at the Department of Statistics, Princeton University. This is an example of "gray literature"; the statistics department was later closed.
4. Zipf, G. K. (1949), Human Behavior and the Principle of Least Effort, Cambridge, Mass.: Addison-Wesley.
5. Carmalt, S., and B. St. John (1986), "Giant Oil and Gas Fields," American Association of Petroleum Geologists Memoir 40, pp. 11–52.
6. Knebel, G. M., and E. G. Rodriguez (1956), "Habitat of Some Oil," American Association of Petroleum Geologists Bulletin **40**:547–61.
7. Tukey, J. W. (1977), Exploratory Data Analysis, Reading, Mass.: Addison-Wesley, p. 590. In more modern lingo, Zipf's law is a fractal with Hausdorff dimension of one.
8. Udden, J. A. (1914), "Mechanical Composition of Clastic Sediments," Geological Society of America Bulletin **25**:655–744.
9. Bell, E. T. (1937), Men of Mathematics, New York: Simon & Schuster, pp. 218–69.
10. Deffeyes, K. S., and I. D. MacGregor (1980), "World Uranium Resources," Scientific American **242**:66–76.
11. Van Bellen, R. C. (1956), "The Stratigraphy of the 'Main Limestone' of the Kirkuk, Bai Hassan, and Qarah Chauq Dagh Structures in North Iraq," Journal of the Institute of Petroleum **42**:233–63.
12. For example, Yergin, Daniel (1991), The Prize, New York: Simon & Schuster.
13. Brown, A. C. (1999), Oil, God, and Gold, Boston: Houghton Mifflin.
14. Pekot, L. J., and G. A. Gersib (1987), "Ekofisk," in Geology of the Norwegian Oil and Gas Fields, ed. A. M. Spencer, London: Graham & Trotman, pp. 73–87.
15. Production statistics are published by the Oil and Gas Journal in the last issue of each year.

———————————— *Chapter 7* ————————————

1. Hubbert, M. K. (1956), "Nuclear Energy and the Fossil Fuels," American Petroleum Institute Drilling and Production Practice, Proceedings of Spring Meeting, San Antonio, 1956, pp. 7–25.
2. Hager, T. (1995), Force of Nature: The Life of Linus Pauling, New York: Simon & Schuster.
3. Doan, D. B. (1994), "Memorial to M. King Hubbert," Geological Society of America Bulletin 24:39–46.
4. U.S. Geological Survey World Energy Assessment Team (2000), "U.S. Geological Survey World Petroleum Assessment 2000," USGS Digital Data Series DDS-60 (4 compact discs).
5. Carmalt, S., and B. St. John (1986), "Giant Oil and Gas Fields," American Association of Petroleum Geologists Memoir 40, pp. 11–52.
6. Tippee, B. (1998), "U.S. Fields with Ultimate Oil Recovery Exceeding 100 Million Barrels," in International Petroleum Encyclopedia, Tulsa: Penn-Well, pp. 321–22.
7. Hubbert ("Nuclear Energy and the Fossil Fuels") credits Wallace Pratt and Lewis Weeks for generating the educated guesses about the U.S. ultimate oil production.
8. Hubbert, M. K. (1962), "Energy Resources," National Academy of Sciences–National Research Council, Publication 1000-D.
9. Claerbout, J. F. (1985), Imaging the Earth's Interior, Palo Alto, Calif.: Blackwell, p. 123.
10. Hubbert, M. K. (1967), "Degree of Advancement of Petroleum Exploration in the United States," American Association of Petroleum Geologists Bulletin 51:2207–27. Because so few people believed him, Hubbert also examined the amount of oil found per foot of exploratory drilling.
11. Obermiller, J. (1999), "Historic World Oil Production," in Basic Petroleum Data Book, Washington, D.C.: American Petroleum Institute, vol. 19, sec, 4, table 10.
12. Campbell, C. J. (1997), The Coming Oil Crisis, Multi-Science Publishing Company and Petroconsultants.
13. Brown, D. (2000), "Bulls and Bears Duel over Supply," American Association of Petroleum Geologists Explorer, May:12–15.
14. Hubbert, M. K. (1981), "The World's Evolving Energy System," American Journal of Physics 49:1007–29.

———————————— *Chapter 8* ————————————

1. Kingsland, S. E. (1985), Modeling Nature, Chicago: University of Chicago Press, a history of ideas in population ecology.
2. Brown, J. (1995), Charles Darwin Voyaging, Princeton: Princeton University Press, p. 385.

3. Verhulst, P. F. (1838), "Notice sur la loi que la population suit dans son acroissement," Corr. Math. et Phys. **10**:113.
4. Hubbert, M. K. (1982), "Techniques of Prediction as Applied to the Production of Oil and Gas," in Oil and Gas Supply Modeling, ed. S. I. Gass, National Bureau of Standards Special Publication 631, pp. 16–141. Here in the back of the book, where my editor isn't likely to look, we can derive the equation:

The logistic equation is:

$$Q = \frac{Q_o}{1 + \exp[a(t_o - t)]} .$$

Hubbert's equation 28

Differentiating with respect to time:

$$dQ/dt = Q_o \frac{a \exp[a(t_o - t)]}{\{1 + \exp[a(t_o - t)]\}^2} .$$

Rearranging the first equation and squaring:

$$\frac{Q^2}{Q_o^2} = \frac{1}{\{1 + \exp[a(t_o - t)]\}^2} .$$

Rearranging the first equation in another way:

$$Q_o/Q - 1 = \exp[a(t_o - t)].$$

Substituting these last two into the time derivative gives

$$dQ/dt = aQ_o \, [(Q_o/Q) - 1][Q/Q_o]^2,$$

which reduces to

$$dQ/dt = a[Q - Q^2/Q_o].$$

Hubbert's equation 24

Dividing both sides by Q gives

$$\frac{dQ/dt}{Q} = a - (a/Q_o)Q.$$

Hubbert's equation 27

The right-hand side of the last equation is a linear expression in Q, which is the basis for the graph.

5. Hubbert, M. K. (1956), "Nuclear Energy and the Fossil Fuels," American Petroleum Institute Drilling and Production Practice, Proceedings of Spring Meeting, San Antonio, 1956, pp. 7–25.

6. Unconstrained exponential growth (compound interest growth):

$dN/dt = rN.$

Logistic population growth:

$$\frac{dN/dt}{N} = \frac{r(K - N)}{K}$$ where $(K - N)/K$ is the empty carrying capacity.

Oil production:

$$\frac{dQ/dt}{Q} = \frac{A(Q_o - Q)}{Q_o}$$ where $(Q_o - Q)/Q_o$ is the undiscovered fraction.

7. Smith, F. E. (1963), "Population Dynamics in Daphnia magna and a New Model for Population Growth," Ecology **44**:651–63. This may be the first use of the growth rate versus population graph.
8. "Probability paper" converts a Gaussian to a straight line but has to express the cumulative production as a fraction of the ultimate final production.
9. Campbell, C. J. (1997), The Coming Oil Crisis, Multi-Science Publishing Company and Petroconsultants, p. 205.
10. U.S. Geological Survey World Energy Assessment Team (2000), "U.S. Geological Survey World Petroleum Assessment 2000," USGS Digital Data Series DDS-60 (4 compact discs).
11. Campbell, "Coming Oil Crisis," p. 73.

———————————— *Chapter 9* ————————————

1. Berry, M. C. (1980), The University of Texas, Austin: University of Texas Press, p. 11. Oil income from 2 million acres in west Texas is divided, two-thirds for UT Austin and one-third for Texas A&M.
2. (1999), "Azerbaijan Signs Three E&P Deals," Oil and Gas Journal, May 3.
3. Chorn, L. G., and M. Croft (2000), "Resolving Reservoir Uncertainty to Create Value," Journal of Petroleum Technology, August:52–59.
4. West, J. (1996), "Optimism Prevails in World Petrochemical Industry," in International Petroleum Encyclopedia, Tulsa: PennWell, pp. 5–24.
5. Reinharz, J. (1993), Chaim Weizmann, Oxford: Oxford University Press, p. 40.
6. Glasscock, C. B. (1938), Then Came Oil, New York: Grosset & Dunlap. *Oil for the Lamps of China* is the title of a book by Alice Tisdale Hobart.
7. Royal Dutch/Shell staff (1983), The Petroleum Handbook, Amsterdam: Elsevier, p. 279.
8. Pauling, L. C. (1964), College Chemistry, 3d ed., San Francisco: W. H. Freeman, p. 406.

9. Tippee, B. (1999), "Saudi Arabia," in International Petroleum Encyclopedia, Tulsa: PennWell, pp. 93–94.

10. Tippee, B. (1999), Active U.S. EOR Projects, in International Petroleum Encyclopedia, Tulsa: PennWell.

11. Moritis, G. (2000), "EOR Weathers Low Oil Prices," Oil and Gas Journal, March 20, 39–61.

12. McCaslin, J. C., ed. (1987), International Petroleum Encyclopedia, Tulsa: PennWell, p. 63.

13. Garrett, D. E. (1992), Natural Soda Ash, New York: Van Nostrand Reinhold.

14. Deffeyes, K. S. (1982), "Geological Estimates of Methane Availability," in Methane, Fuel for the Future, ed. P. McGeer and E. Durbin, New York: Plenum Press, pp. 19–29.

15. "Worldwide Look at Reserves and Production" (2000), Oil and Gas Journal, December 18, 122–23.

—————————— *Chapter 10* ——————————

1. Armstead, H. T. H. (1978), Geothermal Energy, New York: John Wiley, p. 71.

2. Boron was harvested at Lardarello long before it was developed as a geothermal resource. Dickson, M. H., and M. Fanelli (1995), Geothermal Energy, New York: John Wiley, p. 173.

3. Smith, M. C., and G. M. Ponder (1981), "Hot Dry Rock Geothermal Energy Development Program," Los Alamos National Laboratory, Report LA-9287-HDR.

4. International Atomic Energy Agency (1975), Le Phenomene d'Oklo, Vienna: Agence Internaionale de L'cncrgic Atomiquc.

5. Nero, A. V. (1979), Nuclear Reactors, Berkeley: University of California Press, pp. 77–132.

6. Deffeyes, K. S., and I. D. MacGregor (1980), "World Uranium Resources," Scientific American **242:**66–76.

7. Lester, M. D. (1984), "Energy's Influence on the Bauxite Industry," in Bauxite, ed. L. Jacob, New York: American Institute of Mining, Metallurgical, and Petroleum Engineers, pp. 862–69.

—————————— *Chapter 11* ——————————

1. Gary, J. H., and G. E. Handwerk (1994), Petroleum Refining, New York: Marcel Dekker. In particular, see the preface on p. iii.

2. Pitzer, K. S., and L. Brewer (1961), Thermodynamics, New York: McGraw-Hill. This book is a revision of a classic by G. N. Lewis and Merle Randall. Thermodynamics is never easy, but this book is straightforward.

3. Hubbert, M. K. (1956), "Nuclear Energy and the Fossil Fuels," American Petroleum Institute Drilling and Production Practice, Proceedings of

Spring Meeting, San Antonio, 1956, pp. 7–25; also Shell Development Company Publication 95, June 1956.

4. Thomasson, M. R. (2000), "Petroleum Geology, Is There a Future?" American Association of Petroleum Geologists Explorer, May:3–10.

INDEX